Agents and Actions Supplements
Vol. 46

Series Editors
K. Brune, Erlangen
M.J. Parnham, Bonn

Novel Molecular Approaches to Anti-Inflammatory Therapy

Edited by
W. Pruzanski
P. Vadas

Springer Basel AG

Editors:

Waldemar Pruzanski, MD, FRCPC, FACP, FACR
The Wellesley Hospital Research Institute
University of Toronto
160 Wellesley Street East
Toronto
Ontario
Canada M4Y 1J3

Peter Vadas, MD, PhD, FRCPC, FACP
The Wellesley Hospital Research Institute
University of Toronto
160 Wellesley Street East
Toronto
Ontario
Canada M4Y 1J3

A CIP catalogue record for this book is available from the Library of Congress, Washington D.C., USA

Deutsche Bibliothek Cataloging-in-Publication Data
Novel molecular approaches to anti-inflammatory therapy / ed.
by Waldemar Pruzanski ; P. Vadas. - Basel ; Boston ; Berlin :
Birkhäuser, 1995
   (Agents and actions supplements ; Vol. 46)
NE: Pruzanski, Waldemar; Agents and actions / Supplements

Softcover reprint of the hardcover 1st edition 1995
Camera-ready copy prepared by the editors and authors
Printed on acid-free paper produced from chlorine-free pulp
Cover design: Heinz Hiltbrunner, Basel

ISBN 978-3-0348-7278-2   ISBN 978-3-0348-7276-8 (eBook)
DOI 10.1007/978-3-0348-7276-8

Contents

Introduction

In July 1994, the Canadian Inflammation Society in Toronto hosted a symposium entitled "Novel molecular approaches to anti-inflammatory therapy" as a satellite of the XIIth International Congress of Pharmacology. For several reasons the above topic was singled out and separated from the general Congress. Major advances have recently been made in the understanding of inflammatory mechanisms. These advances serve as the basis for a new generation of therapeutic agents which may potentially alter our approach to inflammatory processes and infectious diseases. The Satellite Symposium focused on recent advances in diversified fields including arachidonate metabolites, actions of cytokines, chemokines, cellular adhesion molecules and novel molecular approaches in the therapy of inflammation. Scholarly addresses were presented by eminent experts in their respective fields. In this book are included 17 papers which describe in detail novel approaches to the study of inflammation. We hope that this book will serve not only as a summary of present knowledge, but will stimulate both scientists and clinicians to continue their efforts to battle inflammatory diseases.

W. Pruzanski,
Chair, Publications Committee

P. Vadas,
Chair, Organizing Committee

EOSINOPHIL CHEMOATTRACTANTS GENERATED IN VIVO

T.J.Williams, D.A.Griffiths-Johnson, P.J.Jose and P.D.Collins.

Department of Applied Pharmacology, National Heart and Lung Institute, Dovehouse Street, London SW3 6LY, United Kingdom.

SUMMARY: The eosinophil is the predominant inflammatory cell which accumulates in the asthmatic lung. There is considerable circumstantial evidence linking these cells to lung dysfunction, but the precise cause and effect relationship is controversial. The defensive role of the eosinophil appears to be concerned largely with eliminating helminth parasites which do not normally present a constant threat. Thus, unlike the neutrophil whose defensive role against microbes is essential, the eosinophil presents a target for therapeutic intervention which is potentially applicable to long-term treatment. Several approaches to suppressing eosinophil accumulation are possible, based on the multiple steps involved in their appearance and activation in the lung (for review see [1]). One approach is to block the receptor(s) to the important endogenous eosinophil chemoattractants generated in the asthmatic lung, offering the potential for selective leukocyte-type suppression. A first step in this pursuit is the identification of such chemoattractants. This article describes recent attempts in this direction, with the long-term goal of producing chemoattractant receptor antagonists.

INTRODUCTION

Eosinophil accumulation is a marked feature of local host defence reactions to helminth infection. The eosinophil is equipped to attach to helminths and release toxic constituents in order to kill the parasites. Helminths can stimulate immune responses resulting characteristically in the generation of IgE. Persistent infection or reinfection of a tissue results in a local reaction in which chemical signals, 'chemoattractants', are generated that have the function of recruiting eosinophils from blood microvessels. That selective eosinophil accumulation occurs under these circumstances when, for example, many more neutrophils are in the circulation, implies that eosinophil-selective chemoattractants are involved. The chemoattractant is the signal that triggers a complex sequence of events dependent on interactions between adhesion molecules and their complementary ligands on eosinophils and microvascular endothelial cells (for review see [2]). The first selectin-mediated phase involves loose tethering dependent on carbohydrate interactions. This is followed by firm attachment and migration through endothelial junctions, both dependent on integrin/CAM interactions. Experiments in vitro [3,4] and in vivo [5] imply that the β_1-integrin, VLA-4, is particularly important with respect to eosinophil accumulation.

Chemical signals working over longer range than chemoattractants are also involved in eosinophil recruitment: these stimulate the bone marrow to release more eosinophils and prime those in the circulation (for review see [1]).

EOSINOPHILS IN ASTHMA

Allergic asthma appears to be a perversion of a host defence reaction to helminths. In an inappropriate response to a provoking stimulus, an acute bronchoconstriction resulting from mediators derived from mast cells is followed by a massive cellular infiltrate largely comprising of eosinophils, frequently associated with a late bronchoconstrictor response and hyper-reactivity to spasmogens. There is strong evidence that these events are controlled by CD4[+] T-cells [6,7] and that the presence of eosinophils is linked to lung dysfunction, but it is clear that there are other causal links which may dominate in some models. The numbers of eosinophils in human asthmatic lung correlate with the degree of lung dysfunction [8,9]. In addition, interventions which suppress eosinophil accumulation in animal models of allergic asthma suppress lung dysfunction in parallel [6,10,11]. However, this is an oversimplification as the two events can be separated experimentally under some circumstances [12]

NON-SELECTIVE EOSINOPHIL CHEMOATTRACTANTS

Much of our knowledge concerning leukocyte chemoattractants originates from use of the Boyden chamber which measures chemotaxis in vitro. This has been of enormous value in predicting candidates for endogenous chemoattractants, although clearly it provides a poor analogue of the dynamic events occurring in the microcirculation in vivo. The first chemoattractant for neutrophils demonstrated using this system was the complement fragment C5a. This has subsequently been shown to be a powerful eosinophil chemoattractant in vitro and in vivo [13,14]. However, C5a lacks selectivity as do the more recently discovered mediators, platelet activating factor and leukotriene B_4 (LTB$_4$) which have also been shown to attract eosinophils in vitro and in vivo [14-17]. Evidence has been presented that LTB$_4$ is involved in eosinophil accumulation in allergic reactions in vivo as an LTB$_4$ antagonist has been shown to suppress eosinophil accumulation in the lungs of challenged/sensitised guinea pigs [18] and a 5-lipoxygenase inhibitor has been shown to suppress eosinophil accumulation in passive cutaneous anaphylactic reactions [19].

SELECTIVE EOSINOPHIL CHEMOATTRACTANTS

Interleukin-5 (IL-5) has received considerable attention as a potential eosinophil-specific chemoattractant in allergic reactions. IL-5 was originally discovered because of its effect on eosinophil proliferation in bone marrow [20]. Subsequently, it was shown to be chemotactic for eosinophils in Boyden chambers and to be upregulated in tissues of sensitised individuals after allergen challenge [21]. Further, neutralising antibodies to IL-5 have been shown to suppress eosinophil accumulation and bronchial hyperresponsiveness in animal models of allergic asthma [12]. However, IL-5 is not a potent chemoattractant; intradermal injection of human recombinant material in guinea pigs does not induce the accumulation of eosinophils [22]. It is possible that IL-5 is more important in vivo for its effects on eosinophil proliferation in bone marrow, priming of eosinophils in the circulation and increasing survival of eosinophils in tissues.

As discussed before, the fact that selective eosinophil accumulation occurs in vivo suggests that selective chemoattractants are important in allergic reactions in vivo. Against this background we carried out experiments in an attempt to identify endogenous chemoattractants using an established guinea pig model of allergic asthma [23]. Sensitised guinea pigs were challenged with aerosolised ovalbumin and bronchoaveolar lavage (BAL) fluid collected at intervals. BAL fluid was injected intradermally into unsensitised bioassay guinea pigs, previously injected with ^{111}In eosinophils as described in our earlier studies [14]. Using the assay to provide an index of chemoattractant activity, it was found that such activity was present at 30 minutes after challenge with a peak at 3-6 hours. Little or no activity was found in control animals (figure 1).

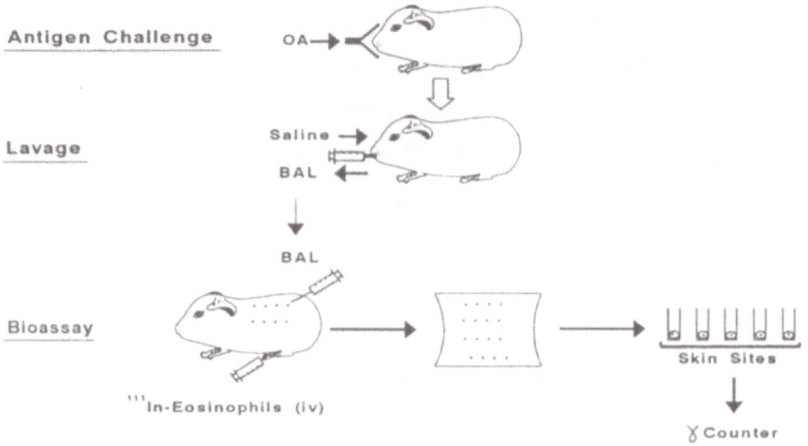

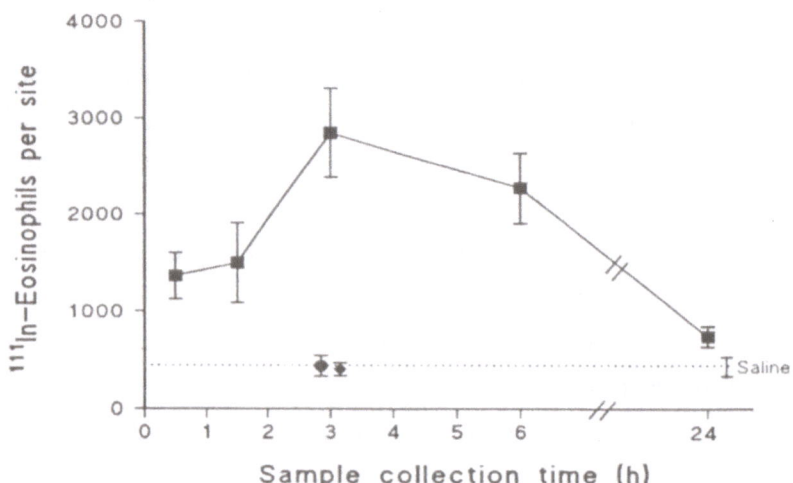

FIGURE 1.   Time course of generation of eosinophil chemoattractant activity in lungs of sensitised guinea-pigs after allergen challenge (■ ,n=4-10).   Activity was measured using an in vivo skin assay of ^{111}In-eosinophil accumulation in unsensitised guinea-pigs over 4h (3 test animals per BAL sample). Responses to control lavage samples obtained 3h after sham (saline) challenge of sensitised animals (♦ ,n=5) or allergen challenge of sham (saline) sensitised animals (● ,n=5) are also shown.

* p < 0.005 compared to responses to intradermal saline, shown as the dotted line.

Taken from Jose et al. J. Exp. Med. *179*, 881-887 (1994) with permission.

Three hour BAL fluid from test animals was purified to homogeneity in a series of HPLC steps, using the skin system to assay fractions throughout. Purified protein appeared in four variants, thought to differ in the extent of glycosylation. Although the purified material was highly potent in inducing eosinophil accumulation in the skin (figure 2) [24] and lung [23] significantly, it was inactive in inducing neutrophil accumulation. The protein was sequenced revealing a novel 73 amino acid member of the C-C family of chemokines which we termed "eotaxin" (figure 3) [24].

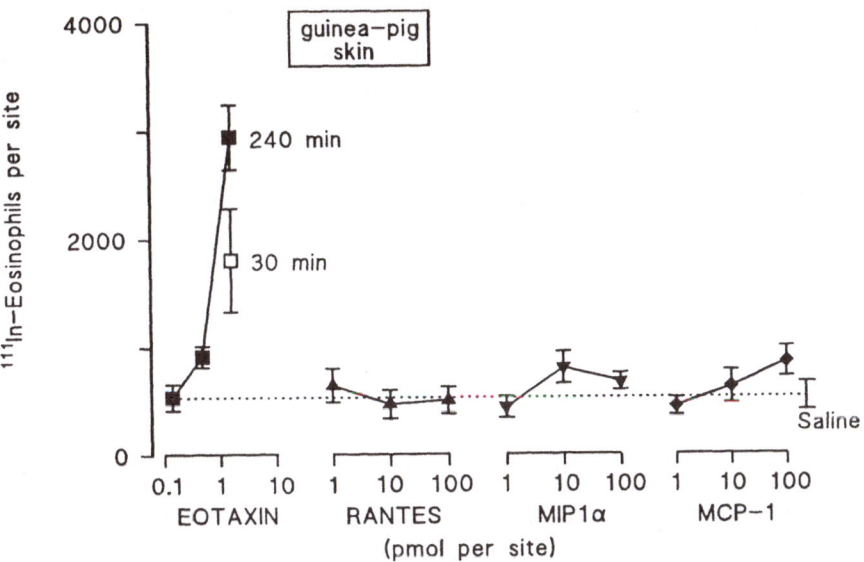

FIGURE 2.   Guinea-pig eotaxin (1.6pmol) induces significant ^{111}In-eosinophil accumulation in vivo 30min (□) and 4h (■) after intradermal injection. In contrast, the recombinant human proteins, RANTES, MIP-1α and MCP-1, at doses up to 100pmol, were without effect over 4 hours. Results are the mean ± sem for n=4 assay animals.

Taken from Jose et al. J. Exp. Med. *179*, 881-887 (1994) with permission.

During the progress of this work, other members of the C-C chemokine family have been tested on eosinophils in vitro. Interestingly, it was shown that platelets could be stimulated to release an eosinophil chemoattractant which was identified as RANTES [25]. Subsequently, other members have been shown to have this activity ie MIP-1$_\alpha$ [26] and most recently MCP-3 [27]. Homology of eotaxin with these human proteins is RANTES 26%,

MIP-1$_\alpha$ 31% and MCP-3 51% (figure 3). In the guinea pig skin system none of these human proteins is active in inducing the accumulation of ^{111}In-eosinophils, whereas eotaxin induces significant responses at 1-2 picomole doses (figure 2). Eotaxin is possibly related to a C-C chemokine expressed in sensitised/challenged mouse mast cells [28], however, homology is not high at 41%.

EOTAXIN

Gp EOTAXIN	HPGIPSACCFRVTNKKI SFQRLKSYKIITSSKCPQTAIVFEIKPDKMICADPKXXWVQDAKKYLDQISQXTKP
Hu MCP-1	QPDAINAPVTCCYNFTNRKIS/QRLASYRRITSSKCPHEAVIFKTIVAKEICADPKQHWVQDSMDHLDHQTQTPKT
Hu MCP-2	DSVSIPITCCFNVINRKIPIQRLESYTPTNICCPHEAVIFKTKRGKEVCADPKFFWVHDSMKHLDQIFQNIKP
Hu MCP-3	KSTTCCYRFINKKIPHQRLESYRRTTSSHCPRFAIFKDKEICADPTQHWVQDFMKHLDHHTQTPKL
Gp MCP-1	GVNTETCCYTFNKQIPLKRVKGYERITSSRCPQEAVIFRTLKNKEVCADPTQKWVQDYIAKTDQRTQQKQN
Hu MIP-1α	SLAADTPTACCFSYTSRQIPQNFIADYFETSSQCSKPGVIFLTKRSRQVCADPSEENVQHYVSDLELSA
Hu MIP-1ß	PMGSDPPTACCFSYTARKLPHNFVVDYYETSSLCSQPAVFQTKRSKQVCADPSESWVQEYVYDLELN
Hu RANTES	SPYSSDTTPCCFAYIARPLPRAHIKEYFYTSGKCSNPAVFVTRKNRQVCANPEKKWVREYINSLEMS

FIGURE 3. The amino acid sequence of eotaxin and comparison with human MCP-1, MCP-2, MCP-3, guinea-pig MCP-1, human MIP-1α, MIP-1ß and RANTES showing conserved residues (shaded).

In vivo, eotaxin induced substantial accumulation of eosinophils over 30 minutes, implying a direct action on eosinophils in vivo. This was supported by results in vitro showing that the chemokine induced mobilisation of intracellular calcium, aggregation and chemotaxis of guinea pig eosinophils. Interestingly, eotaxin induced responses in human eosinophils as well as guinea pig cells. Further, although human RANTES did not stimulate guinea pig eosinophils it appeared to bind to a common receptor with eotaxin.

CONCLUSION

The influence that eosinophils have on lung function in the asthmatic is controversial. One approach to therapy is to block receptors to chemoattractants. We now have the opportunity to evaluate this approach in animals models with the long-term goal of developing selective therapeutic agents.

ACKNOWLEDGEMENTS

We are indebted to the National Asthma Campaign UK and the Wellcome Trust UK for generously supporting our research.

REFERENCES

1] V. B. Weg and T. J. Williams, *Chemical mediators and adhesion molecules involved in eosinophil accumulation in vivo*. Ann. NY Acad. Sci. *in press*, (1994).

2] T. A. Springer, *Traffic signals for lymphocyte recirculation and leukocyte emigration: the multistep paradigm*. Cell *76*, 301-314 (1994).

3] A. Dobrina, R. Menegazzi, T. M. Carlos, E. Nardon, R. Cramer, T. Zacchi, J. M. Harlan and P. Patriarca, *Mechanisms of eosinophil adherence to cultured vascular endothelial cells. Eosinophils bind to the cytokine-induced endothelial ligand vascular cell adhesion molecule-1 via the very late activation antigen-4 integrin receptor*. J. Clin. Invest. *88*, 20-26 (1991).

4] G. M. Walsh, J-J. Mermod, A. Hartnell, A. B. Kay and A. J. Wardlaw, *Human eosinophil, but not neutrophil, adherence to IL-1-stimulated human umbilical vascular endothelial cells is $\alpha_4\beta_1$ (very late antigen-4) dependent*. J. Immunol. *146*, 3419-3423 (1991).

5] V. B. Weg, T. J. Williams, R. R. Lobb and S. Nourshargh, *A monoclonal antibody recognising very late activation antigen-4 (VLA-4) inhibits eosinophil accumulation in vivo*. J. Exp. Med. *177*, 561-566 (1993).

6] S. H. Gavett, X. Chen, F. Finkelman and M. Wills-Karp, *Depletion of murine CD4+ T lymphocytes prevents antigen-induced airway hyperreactivity and pulmonary eosinophilia*. Am. J. Respir. Cell Mol. Biol. *10*, 587-593 (1994).

7] M. Azzawi, B. Bradley, P. K. Jeffery, A. J. Frew, A. J. Wardlaw, G. Knowles, B. Assoufi, J. V. Collins, S. Durham and A. B. Kay, *Identification of activated T lymphocytes and eosinophils in bronchial biopsies in stable atopic asthma*. Am. Rev. Respir. Dis. *142*, 1407-1413 (1990).

8] J. Bousquet, P. Chanez, J. Y. Lacoste, G. Barneon, M. N. Ghavanian, I. Enander, P. Venge, S. Ahlstedt, J. Simony-Lafontaine, P. Godard and P-B. Michel, *Eosinophilic inflammation in asthma*. N. Engl. J. Med. *323*, 1033-1039 (1990).

9] B. L. Bradley, M. Azzawi, M. Jacobson, B. Assoufi, J. V. Collins, A-M. A. Irani, L. B. Schwartz, S. R. Durham, P. K. Jeffery and A. B. Kay, *Eosinophils, T-lymphocytes, mast cells, neutrophils, and macrophages in bronchial biopsy specimens from atopic subjects with asthma: Comparison with biopsy specimens from atopic subjects without asthma and normal control subjects and relationship to bronchial hyperresponsiveness*. J. Allergy Clin. Immunol. *88*, 661-674 (1991).

10] C. D. Wegner, R. H. Gundel, P. Reilly, N. Haynes, L. G. Letts and R. Rothlein, *Intercellular adhesion molecule-1 (ICAM-1) in the pathogenesis of asthma*. Science *247*, 456-459 (1990).

11] A. J. M. Van Oosterhout, R. S. Ladenius, H. F. J. Savelkoul, I. Van Ark, K. C. Delsman and F. P. Nijkamp, *Effect of anti-IL-5 and IL-5 on airway hyperreactivity and eosinophils in guinea pigs*. Am. Rev. Respir. Dis. *147*, 548-552 (1993).

12] P. J. Mauser, A. Pitman, A. Witt, X. Fernandez, J. Zurcher, T. Kung, H. Jones, A. S. Watnick, R. W. Egan, W. Kreutner and K. G. Adams, *Inhibitory effect of the TRFK-5 anti-IL-5 antibody in a guinea pig model of asthma*. Am. Rev. Respir. Dis. *148*, 1623-1627 (1993).

13] A. B. Kay, H. S. Shin and K. F. Austen, *Selective attraction of eosinophils and synergism between eosinophil chemotactic factor of anaphylaxis (ECF-A) and a fragment cleaved from the fifth component of complement (C5a)*. Immunology *24*, 969-976 (1973).

14] L. H. Faccioli, S. Nourshargh, R. Moqbel, F. M. Williams, R. Sehmi, A. B. Kay and T. J. Williams, *The accumulation of ^{111}In-eosinophils induced by inflammatory mediators in vivo*. Immunology *73*, 222-227 (1991).

15] A. J. Wardlaw, R. Moqbel, O. Cromwell and A. B. Kay, *Platelet-activating factor. A potent chemotactic and chemokinetic factor for human eosinophils*. J. Clin. Invest. *78*, 1701-1706 (1986).

16] E. Henocq and B. B. Vargaftig, *Accumulation of eosinophils in response to intracutaneous Paf-acether and allergens in man*. Lancet *i*, 1378-1379 (1986).

17] L. Nagy, T. H. Lee, E. J. Goetzl, W. C. Pickett and A. B. Kay, *Complement receptor enhancement and chemotaxis of human neutrophils and eosinophils by leukotrienes and other lipoxygenase products*. Clin. Exp. Immunol. *47*, 541-547 (1982).

18] I. M. Richards, R. L. Griffin, J. A. Oostveen, J. Morris, D. G. Wishka and C. J. Dunn, *Effect of the selective leukotriene B₄ antagonist U-75302 on antigen-induced bronchopulmonary eosinophilia in sensitized guinea-pigs*. Am. Rev. Respir. Dis. *140*, 1712-1716 (1989).

19] M. M. Teixeira and P. G. Hellewell, *Effect of a 5-lipoxygenase inhibitor, ZM 230487, on cutaneous allergic inflammation in the guinea-pig*. Br. J. Pharmacol. *111*, 1205-1211 (1994).

20] E. J. Clutterbuck and C. J. Sanderson, *Human eosinophil hematopoiesis studied in vitro by means of murine eosinophil differentiation factor (IL-5): Production of functionally active eosinophils from normal human bone marrow*. Blood *71*, 646-651 (1988).

21] J. M. Wang, A. Rambaldi, A. Biondi, Z. G. Chen, C. J. Sanderson and A. Mantovani, *Recombinant human interleukin 5 is a selective eosinophil chemoattractant*. Eur. J. Immunol. *19*, 701-705 (1989).

22] P. D. Collins, V. B. Weg, L. H. Faccioli, M. L. Watson, R. Moqbel and T. J. Williams, *Eosinophil accumulation induced by human interleukin-8 in the guinea-pig in vivo.* Immunology *79*, 312-318 (1993).

23] D. A. Griffiths-Johnson, P. D. Collins, A. G. Rossi, P. J. Jose and T. J. Williams, *The chemokine, eotaxin, activates guinea-pig eosinophils in vitro, and causes their accumulation into the lung in vivo.* Biochem. Biophys. Res. Commun. *197*, 1167-1172 (1993).

24] P. J. Jose, D. A. Griffiths-Johnson, P. D. Collins, D. T. Walsh, R. Moqbel, N. F. Totty, O. Truong, J. J. Hsuan and T. J. Williams, *Eotaxin: A potent eosinophil chemoattractant cytokine detected in a guinea-pig model of allergic airways inflammation.* J. Exp. Med. *179*, 881-887 (1994).

25] Y. Kameyoshi, A. Dorschner, A. I. Mallet, E. Christophers and J-M. Schroder, *Cytokine RANTES released by thrombin-stimulated platetets is a potent attractant for human eosinophils.* J. Exp. Med. *176*, 587-592 (1992).

26] A. Rot, M. Krieger, T. Brunner, S. C. Bischoff, T. J. Schall and C. A. Dahinden, *RANTES and macrophage inflammatory protein 1α induce the migration and activation of normal human eosinophil granulocytes.* J. Exp. Med. *176*, 1489-1495 (1992).

27] C. A. Dahinden, T. Geiser, T. Brunner, V. Von Tscharner, D. Caput, P. Ferrara, A. Minty and M. Baggiolini, *Monocyte chemotactic protein 3 is a most effective basophil- and eosinophil-activating chemokine.* J. Exp. Med. *179*, 751-756 (1994).

28] P. A. Kulmburg, N. E. Huber, B. J. Scheer, M. Wrann and T. Baumruker, *Immunoglobulin E plus antigen challenge induces a novel intercrine/chemokine in mouse mast cells.* J. Exp. Med. *176*, 1773-1778 (1992).

AAS 46
Novel Molecular Approaches
to Anti-Inflammatory Theory
© 1995 Birkhäuser Verlag Basel

CHEMOKINES AND THEIR ROLE IN HUMAN DISEASE

Steven L. Kunkel, Nicholas Lukacs, and Robert M. Strieter*

The University of Michigan Medical School, Departments of Pathology and Internal Medicine, Division of Pulmonary and Critical Care Medicine*, 1301 Catherine Road, Ann Arbor, Michigan 48109-0602, USA

Summary

The recruitment of leukocyte populations to an area of inflammation is one of the most fundamental processes of immune reactivity, yet a number of the mechanisms which are important to this process are not clearly understood. Investigations directed at understanding the mechanisms of leukocyte elicitation have centered around classical chemotactic factors such as C5a and fMLP, however, these known agents have demonstrated little specificity for recruiting particular leukocyte populations. Recent advances in this field have been made with the discovery of a novel supergene family of chemotactic cytokines or chemokines. These cytokines are important as they posess a high degree of specificity for the recruitment of specific subpopulations of leukocytes.

INTRODUCTION

The inflammatory response constitutes the host's reaction to a number of direct and indirect insults, including infections, soft tissue trauma, cancer, and autoimmune disease. The immune response which is established following one of the above disorders is often characterized as an acute or chronic inflammatory reaction, depending upon the longevity of the response and the specific populations of infiltrating leukocytes. The actual mechanisms that direct leukocyte involvement in the initiation, maintenance, and repair process of acute and chronic inflammation is currently under study. It is becoming increasingly clear that both peripheral blood leukocytes and tissue associated leukocytes, in combination with resident structural cells, are all key participants in maintaining cytokine cascades which mobilize subpopulations of leukocytes to an inflamed area. The host's reaction to an infectious process is an example of a concerted effort of various cells to maximize the inflammatory response and recruit the appropriate leukocyte population. The introduction of an infectious agent results in the generation of early response cytokines, such as interleukin-1 (IL-1) and tumor necrosis factor (TNF), from tissue macrophages or circulating leukocytes (1) These early response cytokines can in turn activate surrounding resident tissue cells which respond by generating additional cytokines needed to maintain the response. This cytokine network is important in localizing the area of inflammation, thus, restricting the area of reactivity, and providing an additional source of mediators from the cells which comprise the inflamed tissue. While the injury associated with acute inflammation may be significant, tissue repair and remodeling can occur with the reestablishment of normal tissue function.

One of the strict characteristics of acute and chronic inflammation is the accumulation of specific subpopulations of peripheral blood leukocytes to the inflammatory site. Leukocyte elicitation is associated with an evolving inflammatory response and is dependent upon a complex series of events, including activation of endothelial cells, expression of leukocyte and endothelial cell adhesion molecules, leukocyte transendothelial cell migration, and leukocyte migration along established chemotactic gradients. Within the last decade significant progress has been made toward understanding the mechanisms of leukocyte/endothelial cell interactions. The development of reagents to identify surface adhesion molecules and the discovery of inherited leukocyte adhesion deficiency have provided useful insight into leukocyte-endothelial cell adhesion. In addition, recent advances in understanding leukocyte directed movement have revealed two separate families of novel chemotactic cytokines or chemokines that have significantly advanced our understanding of leukocyte recruitment (2-6). More importantly, information gained from these studies have laid the ground work for the development of new therapeutic strategies to treat both acute and chronic inflammatory diseases.

SUPERGENE FAMILY OF CHEMOTACTIC CYTOKINES

The identification of new chemotactic factors has provided an important impetus to assess leukocyte recruitment at the mechanistic level (4). One of the interesting aspects of the chemokine supergene family is the relative specificity that these polypeptide mediators display for the recruitment of certain peripheral blood leukocyte populations. The importance of this biological activity is exemplified by the redundancy of inflammatory mediators which display leukocyte chemotactic activity. As stated above, the recruitment response is a strict requirement for the normal operation of inflammation, thus the host appears to possess a number of mechanisms which will target leukocytes to an area of immune reactivity. Previous investigations have identified a number of active participants in leukocyte locomotion. These factors represented a wide range of chemical entities, including biologically active lipids, split products of large polypeptides, small polypeptides, and peptides (2-4 amino acids). However, most of these early identified mediators of inflammation lacked specificity for the elicitation of a particular leukocyte population. During the mid-1980s, two supergene families of chemotactic peptides were identified that demonstrated relative specificity for movement and activation of leukocyte populations (6). These chemokines belong to two groups of related polypeptides, identified by the location of two of the four cysteine amino acids comprising their primary amino acid structure.

Accumulating evidence supports the concept that members of these supergene families have pro inflammatory and reparative activities. In their monomeric form these chemokines are less than 10,000 daltons, possess a high isoelectric point, and are basic-heparin binding proteins. One of the chemokine families displays a conserved amino acid motif characterized by the location of two amino terminal cysteines separated by one non conserved amino acid residue. This chemokine family is designated as a C-X-C chemokine and appears to have specificity for the elicitation of neutrophils. The C-X-C chemokines are all clustered on human chromosome 4 and possess approximately 20% to 55% homology in their primary structure. Interest in this area is exemplified by investigations which have identified 12 different C-X-C chemokines, including platelet factor-4, platelet basic protein, connective tissue activating protein III, beta-thromboglobulin, neutrophil activating factor-2, interleukin-8, growth-related oncogene alpha, beta, and gamma, gamma interferon-inducible protein (IP-10), epithelial neutrophil activating protein-78, monokine induced by gamma interferon (MIG), and granulocyte chemotactic protein-2 (GCP-2) (7-13). Connective tissue activating protein-III, beta thromboglobulin, and neutrophil activating protein-2 are all N-terminal truncations of platelet basic protein. These cleavage products are formed when platelet basic protein is released from platelet alpha granules and is proteolytically digested by leukocyte-derived proteases.

Recent investigations have demonstrated that an important feature in the primary structure of C-X-C chemokines may account for the neutrophil chemotactic and activating properties of these mediators (Table 1). These studies identified three critical amino acid residues immediately preceding the first N-terminal cysteine, which are important in binding to a neutrophil receptor and activating neutrophils. These amino acids are Glu-Leu-Arg or the ELR motif, which is absent in certain members of the C-X-C chemokine family (7,8). In particular, platelet factor-4, IP-10, and MIG all lack the ELR motif and do not possess potent neutrophil activating properties. However, when the ELR motif was synthetically introduced into platelet factor-4, this polypeptide gained chemotactic activity. Therefore, certain members of the C-X-C supergene family may have different biological activities. Interestingly, platelet factor-4 (a non-ELR containing C-X-C chemokine) was one of the first members of this family to be described. This factor was originally identified for its ability to bind heparin, leading to the inactivation of the anticoagulation function of heparin. Interleukin-8 has been the most studied C-X-C chemokine and is produced by an array of cells including primary cultures of monocytes, alveolar macrophages, neutrophils, keratinocytes, mesangial cells, epithelial cells, hepatocytes, fibroblasts, and endothelial cells (14-19). In addition, IL-8 is expressed by a number of neoplasms and transformed cell lines.

C-X-C CHEMOKINES CONTAINING THE ELR AMINO ACID MOTIF

-Interleukin-8 (IL-8)
-Growth regulated oncogene alpha (GRO-alpha)
-Growth regulated oncogene beta (GRO-beta)
-Growth regulated oncogene gamma (GRO-gamma)
-Epithelial neutrophil activating protein (ENA-78)
-Neutrophil activating protein-2 (NAP-2)
-Granulocyte chemotactic protein-2 (GCP-2)

C-X-C CHEMOKINES THAT LACK THE ELR AMINO ACID MOTIF

-Monokine induced by gamma interferon (MIG)
-Platelet factor 4 (PF4)
-gamma-interferon inducible protein-10 (IP-10)

Table 1. The C-X-C chemokine family include cytokines which possess or lack the amino terminal motif of glutamic acid, leucine, and arginine.

A family of polypeptide mediators which is related to the C-X-C chemokines is the C-C chemokine family. This group is defined by the location of the first two N-terminal cysteines being found in juxtaposition to one another. In general terms, the C-C chemokine family has relative specificity for the elicitation of mononuclear cells. The C-C chemokine family includes macrophage inflammatory protein-1 alpha and beta, monocyte chemoattractant protein 1,2, and 3, I-309, and RANTES. Like the C-X-C family, certain members of the C-C members have been identified as products from a number of cellular sources, including a variety of tumor cell lines.

CHEMOKINE EXPRESSION BY NON-INFLAMMATORY CELLS

The expression of various chemokines by noninflammatory cells raises an intriguing question regarding the specific role that resident tissue cells play during the initiation and maintenance of an inflammatory response. Historically, investigations have suggested that monokines and lymphokines were only expressed by monocytes and lymphocytes, however, this does not accurately reflect the cell specific cytokine production pattern during inflammation. This is especially true regarding the production of chemokines, as stromal cells, epithelial cells, endothelial cells, and hepatocytes can all express significant levels of these polypeptide mediators of inflammation. These studies support the notion that normal resident tissue cells can support an inflammatory response via the induction of appropriate chemokines. The production of either C-C or C-X-C family members by local resident cells is dependent upon a cytokine network. In this network or cascade, early response cytokines or master cytokines, such as IL-1 or TNF, can dictate the expression of subsequent chemokines (20,21). This communication process is likely controlled at the tissue level by fixed, resident macrophages. These sentinel macrophages can respond rapidly to injury or a foreign challenge and release IL-1 and/or TNF, which can activate the surrounding resident tissue cells in a cytokine specific manner (Table 2).

| | STIMULUS | | | |
CELLULAR SOURCE	LPS	TNF	IL-1	IL-6
Epithelial Cells	---	++	+++	---
Fibroblasts	+/-	++	+++	---
Endothelial Cells	++	++	+++	---
Monocytes	+++	++	+++	---
Alveolar Macrophages	+++	++	+++	---

Table 2. IL-8 is expressed by a variety of cells in a stimulus specific manner.

Resident non-immune cells are usually not susceptible to a challenge with specific inflammatory activating agents like lipopolysaccharide or immune complexes. Therefore, the production of IL-1 or TNF by local resident macrophages can magnify the response needed to initiate and maintain the recruitment process. These networks have been assessed in the context of both the lung and liver, where early response cytokines released by alveolar macrophages or Kupffer cells can interact with surrounding type II alveolar epithelial cells or hepatocytes, respectively. This communication cascade has been demonstrated in vitro using either co-culture systems or non-immune cells stimulated with macrophage conditioned media. In co-culture experiments, alveolar macrophages were cultured together with lung fibroblasts in the presence of LPS and assessed for the time-dependent production of IL-8. Within 2 hours post LPS addition, alveolar macrophages, but not lung fibroblasts were expressing antigenic IL-8, as determined by immunolocalization. By 8-12 hours, both cells types were producing IL-8 antigen. Interestingly, the addition of neutralizing antibody to TNF or IL-1 to the co-culture dramatically suppressed the production of IL-8 by the lung fibroblast at all time points. However, the expression of IL-8 by the alveolar macrophages was not effected by the antibody treatment during the times studied. Subsequent studies have also demonstrated the production of C-C chemokines by non-immune cells is regulated in a similar manner. However, there appears to be an interesting difference between the induction of C-C and C-X-C members driven by LPS-induced cytokine networks. Lipopolysaccharide is a potent inducing signal for the expression of IL-8,but not MCP-1, from alveolar macrophages and Kupffer cells. Taken in total, this communication network supports the inflammatory response by increasing the number of participating effector cells involved in leukocyte recruitment.

THE ROLE OF IL-8 IN PULMONARY DISEASE

Initial experimental animal studies attempting to assess the in vivo activity of IL-8 demonstrated that a single 100 ug intravenous injection of into rabbits caused a marked neutrophilia, which was observed after 15 minutes and peaked by 1-2 hours post challenge (22). Histological examination revealed both leukostasis and congestion in dilated lung vessels in the IL-8 treated animals. Interestingly, repeated intravenous administration of IL-8 induced significant histological alterations, which appeared to be restricted to the lungs. These changes included the formation of neutrophil aggregates in the small to medium-sized lung vessels, diffuse septal and intra-alveolar edema, and foci of hemorrhage. Animals that received 5 daily injections (5nmol each) of IL-8 showed a substantial broadening of the alveolar septa, which contained fibroblasts and a mixed leukocytic infiltrate. This information supports the earlier studies demonstrating that the intradermal administration of IL-8 in experimental animals can induce leukocyte accumulation in the skin. Furthermore it provides interesting information concerning the lung as a sensitive in vivo target for the pathologic effects of IL-8 and likely other C-X-C chemokines. IL-8 has also been identified as an important mediator in additional models of experimental lung inflammation. In a rabbit model of lung reperfusion injury, antigenic IL-8 was observed both in the BAL fluid and in lung tissue homogenates after 2 hours of ischemia followed by 2 hours of reperfusion. The levels of IL-8 correlated with neutrophil infiltration and destruction of the pulmonary architecture (23). The administration of neutralizing IL-8 antibody prevented neutrophil infiltration and lung injury, suggesting a causal role for IL-8 in this model. Furthermore, the initiation of endotoxin-induced pleurisy in the rabbit was associated with a significant increase in IL-8 expression in the pleural space, which was associated with a profound influx of neutrophils. Passive immunization of rabbits with neutralizing antibodies to rabbit IL-8 resulted in a marked reduction in neutrophil influx into the pleural space (24). These studies were important in that species specific reagents were used to confirm the role of IL-8 in an experimental model of acute lung inflammation, since previous reports had demonstrated some efficacy of antibodies to human IL-8 in a rodent model of acute lung injury (25).

INTERLEUKIN-8 IN ADULT RESPIRATORY DISTRESS SYNDROME

Our laboratory and others have identified IL-8 as an important cytokine in the pathogenesis of a number of human diseases, including disorders of the lung. In a recent investigation, we have demonstrated that the early appearance of IL-8 in the bronchoalveolar lavage (BAL) of patients at risk of developing ARDS may be an important prognostic indicator for the clinical development of this syndrome (26). It is generally believed that a significant contributing factor to the pathology of ARDS is the sequestration, emigration, and activation of neutrophils. Although ARDS has been described in patients with peripheral blood neutropenia, there is compelling evidence for the participation of neutrophils in most cases of ARDS. In our studies, high concentrations of IL-8 were found in the BAL from trauma patients, some within 1 hour of injury. Patients who progressed to ARDS had significantly higher levels of IL-8 in the BAL than those patients who did not develop ARDS. One of the cellular sources of IL-8 in these ARDS patients were the alveolar macrophages, however, it is likely other cells in the alveolar/capillary wall may also contribute to the elevated pulmonary levels of IL-8. An important aspect of these studies was the identification of elevated IL-8 levels which preceded the presence of increased neutrophil numbers in the BAL. While an earlier study also identified IL-8 in the BAL fluid of both ARDS patients and patients at-risk of developing ARDS, a significant correlation with mortality was not found (27). However, studies by Miller and colleagues identified elevated levels of IL-8 in the BAL of ARDS patients, as compared to controls, and the IL-8 levels significantly correlated with patient mortality (28). These investigators also demonstrated a significant correlation between the percentage of neutrophils and the IL-8 concentration in the BAL fluid. Furthermore, the IL-8 levels in the lavage fluid was near the optimal concentration necessary to induce in vitro neutrophil chemotaxis.

INTERLEUKIN-8 IN IDIOPATHIC PULMONARY FIBROSIS

An additional pulmonary disease which results in the recruitment of substantial numbers of leukocytes into the pulmonary interstitium is idiopathic pulmonary fibrosis (IPF). Normal differential cell counts from the bronchoalveolar lavage of patients without pulmonary disease is less than 1% neutrophils. However, the alveolitis associated with IPF demonstrates a histopathologic picture of significant neutrophil recruitment into the pulmonary interstitium and airspace. The differential of cells recovered by bronchoalveolar lavage from IPF patients is often greater than 40% neutrophils, while the percentage of other inflammatory cells, such as T-cells, B-cells, and alveolar macrophages remain normal. This suggests that specific chemoattractants for neutrophils are likely responsible for the recruitment and activation of this leukocyte subpopulation. Recent studies have shown that bronchoalveolar lavage fluid recovered from IPF patients contains elevated levels of IL-8 which was associated with high neutrophil numbers. The alveolar macrophage has been one of the fixed tissue macrophages incriminated as a primary source of neutrophil chemoattractants, including IL-8. In both asbestosis and IPF the alveolar macrophage has clearly been identified as a source of potent neutrophil chemoattractants. This observation is of particular interest, since both of these pulmonary disorders are typically characterized by a predominant neutrophil alveolitis. Thus, a common mechanism may be responsible for the recruitment and activation of neutrophils in these disparate clinical states.

INTERLEUKIN-8 IN CYSTIC FIBROSIS

The pathology of bacterial pneumonia is characterized by the accumulation of interstitial and intra-alveolar neutrophils. Interestingly, the histopathology of bacterial bronchopulmonary inflammation in cystic fibrosis patients is characterized by neutrophil infiltration in the terminal bronchioles. To determine the potential role of IL-8 in this disease, the bronchoalveolar lavage of cystic fibrosis patients with and without bronchopulmonary inflammation were assessed for the presence of IL-8. Constitutive levels of steady-state mRNA for IL-8 were found in cystic fibrosis patients with disease exacerbation, whereas IL-8 mRNA levels were not detectable in the lavage from normal volunteers. IL-8 levels, as determined by ELISA, were readily detected in the bronchoalveolar lavage fluid from patients with cystic fibrosis ranging from 332 to 12,900 pg/ml. These IL-8 levels correlated with elevated levels of recruited neutrophils. The IL-8 levels appears to be compartmentalized to the respiratory tract, as IL-8 was not identified in the plasma of these cystic fibrosis patients. Previously C5a was identified as a chemoattractant found in the bronchoalveolar lavage of cystic fibrosis patients, however, the chemotactic activity of specimens pretreated with neutralizing antibodies to IL-8 was reduced by more than 32% when assayed in a modified Boyden chamber. Immunolocalization of antigenic IL-8 demonstrated that the alveolar macrophages and neutrophils recovered by bronchoalveolar lavage were the major sources of this important chemokine. Additional studies were undertaken to establish whether IL-8 expression correlated with disease activity and subsequent response to therapy. A longitudinal study demonstrated the steady state levels of IL-8 mRNA were significantly enhanced in patients with exacerbation of airway inflammation, however, post-antibiotic treatment the IL-8 mRNA levels dropped precipitously. The above findings were interesting in that they provided evidence that IL-8 is one of the major chemotactic factors found in the bronchoalveolar lavage fluid of patients with cystic fibrosis and the levels of IL-8 appeared to correlate with airway inflammation and subsequent therapy.

CONCLUSION

The expression and regulation of chemotactic factors is one of the essential elements for the initiation and maintenance of inflammation. Historically, chemotactic factors such as C5a, fMLP, and LTB4 were defined as major mediators of neutrophil elicitation. However, these factors demonstrated little specificity for the recruitment of subpopulations of leukocytes. The lack of specificity of these classical chemotactic factors was an enigma, as the histopathology of disease demonstrated that certain disorders were characterized by the accumulation of specific subpopulations of leukocytes. The discovery of a novel class of chemotactic cytokines (chemokines) has provided insight into the elicitation of leukocytes, as these inflammatory mediators possess specificity for the recruitment of leukocytes to an area of inflammation. Recent observations have provided evidence that chemokines play an important role in a variety of human diseases. The chemokines designated as C-X-C chemokines possess chemotactic and activating properties mainly for neutrophils, while the C-C chemokines are chemotactic for mononuclear leukocytes. There is little doubt that the development of therapeutics strategies to modulate the expression and/or activity of these chemokines will provide useful treatment modalities for acute and chronic inflammatory disease.

ACKNOWLEDGMENTS

This work was support in part by National Institute of Health grants HL02401, HL44281, HL31693, and HL35276.

REFERENCES

1.   Cybulsky, MI, Chan, MKW, Movat HZ. Acute inflammation and microthrombosis induced by endotoxin, interleukin-1, and tumor necrosis factor, and their implication in gram-negative infections. Lab Invest 1988; 58: 365-378.

2.   Matsushima K, Oppenheim, JJ. Interleukin-8 and MCAF: Novel inflammatory cytokines inducible by IL-1 and TNF. Cytokine 1989; 1: 2-13.

3.   Baggiolini, M, Walz, A, Kunkel SL, Neutrophil-activating peptide-1/interleukin-8, a novel cytokine that activates neutrophils. J Clin Invest 1989; 84: 1045-1049.

4.   Oppenheim, JJ, Zachariae, OC, Mukaida, N, Matsushima, K. Properties of the novel proinflammatory supergene "intercrine" cytokine family. Ann Rev Immunol 1991; 9: 617-648.

5.   Yoshimura, T, Matsushima, K, Tanaka, S, Appella, E, Leonard, E, Oppenheim, JJ. Purification of a human monocyte-derived neutrophil chemotactic factor that has peptide sequence similarity to host defense cytokines. Proc Natl Acad Sci USA 1987; 84: 9233-9237.

6.   Matsushima, K, Morishita, K, Yoshimura, T, Leonard, E, Oppenheim, JJ. Molecular cloning of a human monocyte-derived neutrophil chemotactic factor (MDNCF) and the induction of MDNCF mRNA by interleukin-1 and tumor necrosis factor. J Exp Med 1988; 167: 1883-1893.

7.   Farber JM. HuMIG: a new member of the chemokine family of cytokines. Biochem Biophys Res Comm 1993; 192: 223-230.

8.   Proost P, De Wolf-Peeters C, Conings R, Opdenakker G, Billiau A, Van Damme J. Identification of a novel granulocyte chemotactic protein (GCP-1) from human tumor cells: in vitro and in vivo comparison with natural forms of GROa, IP-10, and IL-8. J Immunol 1993; 150: 1000-1010.

9.   Kaplan G, Luster AD, Hancock G, Cohn Z. The expression of a gamma-interferon-induced protein (IP-10) in delayed immune responses in human skin. J Exp Med 1987; 166:1098-1108.

10.   Ansiowicz A, Zajchowski D, Stenman G, Sager R. Functional diversity of gro gene expression in human fibroblasts and mammary epithelial cells. Proc. Natl. Acad. Sci. USA 1988; 85: 9645-9649.

11.   Ansiowicz A, Bardwell L, Sager R. Constitutive overexpression of a growth-regulated gene in transformed Chinese hamster and human cells. Proc. Natl. Acad. Sci. USA 1987; 84: 7188-7192.

12.   Richmond A, Thomas HG. Melanoma growth stimulatory activity: isolation from human melanoma tumors and characterization of tissue distribution. J. Cell Biochem. 1988; 36: 185-98.

13.   Walz, A, Burgener, R, Car, B, Baggiolini, M, Kunkel, SL, Strieter, RM.
       Structure and neutrophil-activating properties of a novel inflammatory peptide
       (ENA-78) with homology to interleukin-8. J Exp Med 1991; 174: 1355-1362.

14.   Strieter RM, Kunkel SL, Showell H, Remick DG, Phan SH, Ward PA, Marks
       RM. Endothelial cell gene expression of a neutrophil chemotactic factor by TNF,
       LPS, and IL-1. Science 1989; 243:1467-1469.

15.   Strieter RM, Phan SH, Showell HJ, Remick DG, Lynch JP, Genard M, Raiford C,
       Eskandari M, Marks RM, Kunkel SL. Monokine-induced neutrophil chemotactic
       factor gene expression in human fibroblasts. J Biol Chem 1989; 264:10621-10626.

16.   Thornton AJ, Strieter RM, Lindley I, Baggiolini M, Kunkel SL. Cytokine-induced
       gene expression of a neutrophil chemotactic factor/interleukin-8 by human
       hepatocytes. J Immunol 1990; 144: 2609-2613.

17.   Elner VM, Strieter RM, Elner SG, Baggiolini M, Lindley I, Kunkel SL.
       Neutrophil chemotactic factor (IL-8) gene expression by cytokine-treated retinal
       pigment epithelial cells. Am J Path 1990; 136: 745-750.

18.   Strieter RM, Chensue SW, Basha MA, Standiford TJ, Lynch JP, Kunkel SL.
       Human alveolar macrophage gene expression of interleukin-8 by TNF, LPS and
       IL-1. Am J Respir Cell Mol Biol 1990; 2: 321-326.

19.   Brown Z, Strieter RM, Chensue SW, Ceska P, Lindley I, Nield GH, Kunkel SL,
       Westwick J. Cytokine activated human mesangial cells generate the neutrophil
       chemoattractant - interleukin 8. Kidney International 1991; 40: 86-90.

20.   Rolfe MW, Kunkel SL, Standiford TJ, Chensue SW, Allen RM, Evanoff HL,
       Phan SH, Strieter RM. Pulmonary fibroblast expression of interleukin-8: a model
       for alveolar macrophage-derived cytokine networking. Am J Respir Cell Mol Biol
       1991; 5: 493-501.

21   Standiford TJ, Kunkel SL, Basha MA, Chensue SW, Lynch JP, Toews GB,
       Strieter RM. Interleukin-8 gene expression by a pulmonary epithelial cell line: A
       model for cytokine networks in the lung. J Clin Invest 1990; 86:1945-1953.

22.   Zwahlen, R, Walz, A, Rot, A. In vitro and in vivo activity and pathophysiology of
       human IL-8 and related peptides In International Review of Experimental
       pathology. R Zwahlen editor New York Academic Press, 1993: 27-42.

23.   Sekido, N, Mukaida, N, Harada, A, Nakanishi, I, Watanabe, Y, Matsushima, K.
       Prevention of lung reperfusion injury in rabbits by a monoclonal antibody against
       interleukin-8. Nature 1994; 365: 654-657.

24.   Antony, VB, Godbey, SW, Kunkel, SL, Hott, JW, Hartman, Burdick, MD,
       Strieter, RM. Recruitment of inflammatory cells to the pleural space. J Immunol
       1993; 151: 7216-7223.

25.   Mulligan, MS, Jones, ML, Bolanowski, MA, Baganoff, MP, Deppeler,
       CL,Meyers, DM, Ryan, US, Ward, PA. Inhibition of lung inflammatory reactions
       in rats by an anti-human IL-8 antibody. J Immunol 1993; 150: 5585-5592.

26.   Donnelly, SC,Strieter, RM, Kunkel, SL, Walz, A, Robertson, CR, Carter, DC,
       Grant, IS, Pollok, AJ, Haslett, C. Interleukin-8 and development of adult
       respiratory distress syndrome in at-risk patient groups. Lancet 1993; 341: 643-647.

27.   Jorens, PG, Van Dame, J, DeBecker, W, Bossart, L, DeJongh, RF, Herman, AG,
       Rampert, M. Interleukin-8 in the bronchoalveolar lavage fluid from patients with
       the adult respiratory distress syndrome (ARDS) and patients at risk for ARDS.
       Cytokine 1992; 4: 592-597.

28.   Miller, EJ, Cohen, AB, Nagao, S, Griffith, D, Maunder, RJ, Martin, TR, Wiener-
       Kronish, JP, Sticherling, M, Christophers, E, Matthay, MA. Elevated levels of
       NAP-1/interleukin-8 are preseent in the airspaces of patients with the adult
       respiratory distress syndrome and are associated with increased mortality. Am Rev
       Respir Dis 1992; 146: 427-436.

AAS 46
Novel Molecular Approaches
to Anti-Inflammatory Theory
© 1995 Birkhäuser Verlag Basel

STRUCTURE/FUNCTION ANALYSIS OF HUMAN INTERLEUKIN 5 AND ITS RECEPTOR

J. Tavernier, S. Cornelis, R. Devos, Y. Guisez, G. Plaetinck and J. Van der Heyden

Roche Research Gent, Jozef Plateaustraat 22, B-9000 Gent, Belgium

SUMMARY: We have performed a detailed structure-function analysis of human interleukin 5 (hIL5) and its receptor. By testing a hIL5 mutant panel in a solid phase binding assay and a proliferation assay using hIL5 dependent cell-lines, areas on hIL5 involved in either the receptor α-subunit interaction or in receptor activation were identified. Epitope mapping data of a neutralizing and a non-neutralizing monoclonal antibody were in agreement with the mutant analysis. hIL5 binding areas on the IL5Rα-subunit were identfied by interspecies chimaera analysis. Finally, hIL5 mutants with reduced receptor activation potential have antagonistic properties.

INTRODUCTION

On the basis of structural considerations, cytokines can be subdivided into different groups. One class, which comprises most interleukins and colony stimulating factors, has a very conserved four-α-helical bundle structure. Others have an all-β strand scaffold (e.g. IL-1, TNF), or adopt a structure resembling the MHC-family (IL-8 and likely most other chemokines). Such structural relationship is also found for their cognate receptors : the interleukin/haematopoietin receptor superfamily (and the related interferon receptor family) is characterized by having one transmembrane region and a very typical extracellular domain structure. Much in contrast, the interleukin 8 receptor has seven membrane spanning domains, and is related to the rhodopsin receptor family. Also, the members of the TNF superfamily have structurally related receptors. Insight in the structure-activity relationships of cytokines and their receptors is growing rapidly, and is now also revealing conserved interaction patterns.

IL5 is a disulphide-linked homodimeric glycoprotein with 115 amino acids per monomer (1) and has been produced in various heterologous systems (2,3,4). Analysis of the glycosylation pattern of Chinese Hamster Ovary (CHO) cell-derived IL5 revealed an antiparallel dimer linkage. Also, hIL5 was found to have O-linked glycosylation at Thr-3 and N-linked glycosylation at Asn-28 (5). Deglycosylation of IL5 does however not affect its biological activity (2) Very recently, the structure of IL5 purified from *E. coli* (6) and Sf9 cells (7, Oefner et al., personal communication) has been determined. hIL5 adopts the typical four α-helical bundle "cytokine fold", which has also been described for other cytokines including granulocyte-macrophage colony-stimulating factor (GM-CSF, Refs. 8,9), IL2 (10,11), IL4 (12), macrophage colony-stimulating factor (M-CSF, Ref. 13), and growth hormone (GH, Refs. 14,15). This fold consists of a four α-helical bundle in an up-up, down-down array. Unique to IL5 however is the phenomenon of D helix swapping, whereby one bundle is built up of three helices coming from one monomer and a fourth helix which is contributed by the second monomer. In addition to the four α-helical bundle, hIL5 also contains two short antiparallel β-strands located between helices A and B, and C and D respectively.

The human as well as the mouse IL5 receptor consists of two different chains : the α- and β-subunits (16-20). hIL5 binds to the α-subunit with intermediate affinity (Kd $=4.10^{-10}$ M; equivalent to the low affinity binding site in the mouse), which is increased 2-3 fold upon association with the β-chain. This β-subunit is shared with GM-CSF and IL3, which explains the overlap of biological activities observed for these cytokines (16,21,22). Accordingly, cell-type specific expression of the α-subunit (i.e. eosinophils and basophils in the case of hIL5) restricts the activity repertoire to that cell type. We previously reported that mature eosinophils express predominantly a mRNA encoding a soluble receptor α-subunit variant, which has antagonistic properties in vitro (16). This, and another, very similar, minor variant arise from splicing variation (23,24). Remarkably, one hIL5 homodimer binds only one soluble isoform in solution, (25).

MATERIALS AND METHODS

Site-specific mutagenesis and expression

Mutations were introduced in the hIL5 polypeptide by an adapted protocol using a commercially available kit (Transformer, Clontech). Plasmid DNAs harbouring the mutant IL5 genes were introduced in the AcNPV genome by cotransfection with linearized baculovirus DNA in Sf9 cells (Baculogold, Pharmingen). Recombinant baculovirus amplification, mutant hIL5 production and ^{35}S-methionine labeling of hIL5 mutants was essentially performed as described (2). Quantification was performed using hIL5-specific ELISAs.

Analysis of receptor interactions of hIL5 mutants.

The effect of mutations on hIL5Rα binding was monitored seperately on a solid phase binding assay. Essentially, Probind plates (Becton-Dickinson) were coated with polyclonal goat anti hIgG and loaded with shIL5Rα-hIgG3 fusion protein. Then a competition assay was performed by applying serial dilutions of Sf9 supernatants containing hIL5 muteins together with a fixed amount of ^{125}I-hIL5. All mutants were also tested in two different bioassays. First, proliferative activity was monitored using hIL5-dependent FDCP1-CA1 cells. In most cases, hIL5 muteins were also tested on TF1-hIL5Rα cells. Cells were seeded in the presence of serial dilutions of hIL5 (mutant) polypeptides at 1000 or 3000 cells/well for FDCP1-CA1 and TF1-hIL5Rα cells respectively, and proliferation was measured after 65 hours by ^{3}H-Thymidine incorporation for 4-5 hours. Scatchard plot analysis was performed as described before (16).

RESULTS AND DISCUSSION

INTERLEUKIN 5

The structure of human IL5 as deduced from X-Ray crystallographic data is shown in Fig.1.

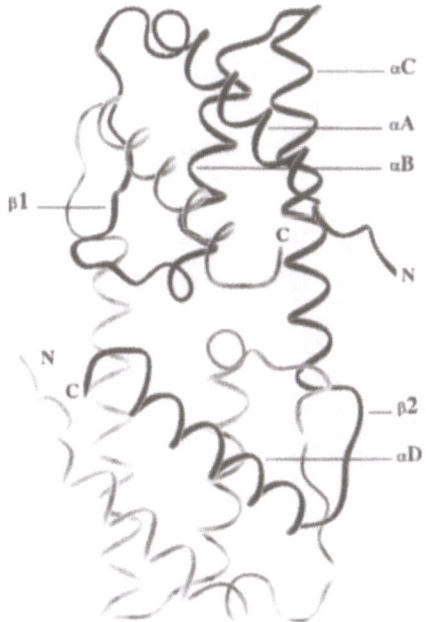

FIG.1., Three-dimensional structure of human IL5,
The human IL5 dimer is presented in full ribbon display, each monomer in two different tones. All α-helices (in the typical up-up, down-down array) and 2 β-strands are indicated, together with the N- and C-termini.

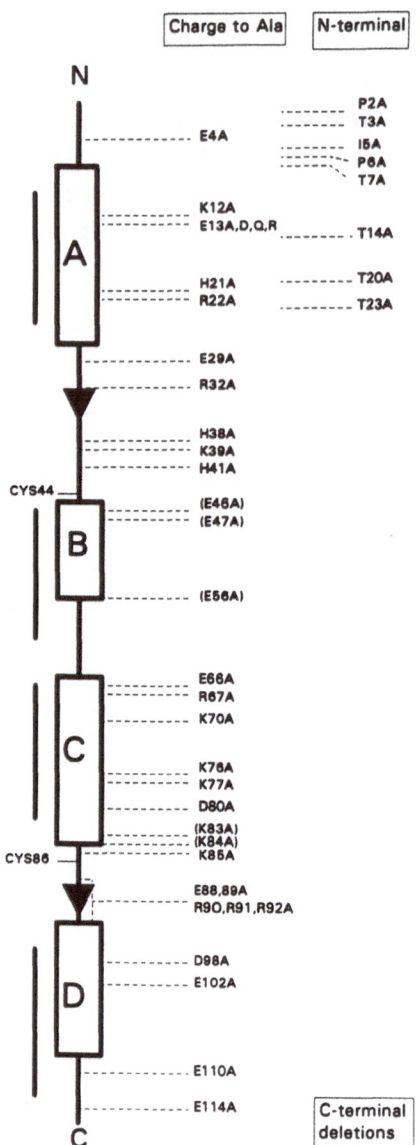

FIG.2., Summary of the human IL5 mutants

An IL5 monomer is schematically shown, with boxes and arrows representing α-helices and β-strands respectively. N and C indicate the orientation. Mutations are presented by their location, with the original and mutant residue at the left and right, respectively. All amino acids are shown in one letter notation.

A series of IL5 mutants was generated directly in pVL941-hIL5, a vector yielding high level expression in Sf9 cells (2). These mutations include an alanine scan of almost all charged residues, and of an additional series of N-terminal residues located at the surface of α-helix A, and a set of C-terminal deletions. All mutations are presented in figure 2. The expression was monitored by internal labeling using ^{35}S-methionine, followed by immunoprecipitation with a polyclonal antiserum against hIL5 and SDS-PAGE analysis. In addition, more accurate quantification was performed using two different ELISAs (see below). Only mutant T3A could not be expressed.

hIL-5 MUTANT		5A5/5A5	H30/POLY
P2A		+ + +	+ + +
T3A			
E4A		+ + +	+ + +
I5A		+ + +	+ + +
P6A		+ + +	+ + +
T7A		+ + +	+ + +
K12A		+ + +	+ + +
E13A		+ + +	+ + +
T14A		+ + +	+ + +
T20A		+ + +	+ + +
H21A		+ + +	+ + +
R22A		+ + +	+ + +
T23A		+ + +	+ + +
E29A		+ + +	+ + +
R32A		0	0
H38A		+ + +	+ + +
K39A		+ + +	+ + +
H41A		+ + +	+ + +
E66A		+ + +	+ + +
R67A		+ + +	+ + +
K70A		+ + +	+ + +
K76A		+ + +	+ + +
K77A		+ + +	+ + +
D80A		+ + +	+ + +
K85A		+ + +	+ + +
E88A		+ + +	+ + +
E89A		0	+ + +
R90A		0	0
R91A		0	+
R92A		+ + +	0
D98A		+ + +	+ + +
E102A		+ + +	+ + +
E110A		+ + +	+ + +
E114A		+ + +	+ + +

FIG.3., Epitope mapping of mABs 5A5 and H30 using human IL5 specific ELISAs
A summary of results from two different ELISAs is shown. Immunoplates were coated with the two mABs. In the first ELISA, detection was using labelled 5A5, taking advantage of the dimeric structure of hIL5. In the second ELISA, detection was with a sandwich using polyclonal antiserum followed by labelled secondary antibody. + + + indicates binding comparable to wild-type hIL5; + indicates binding less than 10% vs. wild-type hIL5; 0 indicates complete loss of binding.

Two monoclonal antibodies (mABs) raised against hIL5 were selected using standard procedures. 5A5 and H30 are respectively a neutralizing and a non-neutralizing mAB. H30 binds stably to the IL5/IL5Rα complex, and can be used to purify this complex (7). Each mAB was separately used in two different ELISAs, allowing us to map their epitopes using our panel of hIL5 mutants. Results are presented in figure 3. Remarkably, both mABs have overlapping epitopes and bind to residues building up the antiparallel β-sheet. However, the 5A5 binding site is located more to the central axis of the IL5 dimer, suggesting that the IL5Rα interaction footprint might overlap the IL5 dimer interface.

The activity of all mutants was evaluated in two different assay systems : first, a human IL5 specific solid phase binding assay (25), and second, a proliferation assay using a hIL5-dependent cell line (FDCP-1-CA1, a FDCP-1 subclone transfected with the human IL5Rα subunit). Mutants with reduced IL5Rα binding were found to score in both assay systems, whilst mutants deficient in the IL5Rβ interaction only exerted a reduced proliferative activity, being fully potent in the solid phase binding assay. As both assay systems are performed under very different conditions (3 hours at room temperature vs. 2-3 days at 37°C) stability control experiments were also included (data not shown). Mutations affecting α-subunit interaction mapped to the loop between β-strand 1 and α-helix B, part of the β-strand 2, and to the C-terminal region. As a consequence of the symmetrical structure of IL5, it is unclear at present whether the α-receptor binding site is formed by residues on one or both monomers. In both cases however, the areas identified are located close to the central axis of the hIL5 dimer, and hence are compatible with the epitope mapping analysis mentioned above. Our screen allowed us to identify one residue involved in β-subunit interaction : E13. An alanine mutation at this position caused a reduction in the proliferation assay to about 5% of wild-type hIL5, whilst binding in the solid phase assay remained unaffected. The identification of position 13 being involved in β-chain interaction agrees well with previous reports where homologous residues were identified in GM-CSF (E21, Refs. 26,27) and IL3 (E22, Ref. 28). Intriguingly, a negatively charged residue at a homologous position appears to be highly conserved in this cytokine family.

THE INTERLEUKIN 5 RECEPTOR α-SUBUNIT

A schematic presentation of the receptor complex is shown in figure 4. The IL5Rα and -β subunits contain three and four so-called "cytokine-receptor submodules" respectively. The modular structure of these receptors is reflected in the organization of their genes (Refs. 29, 24).

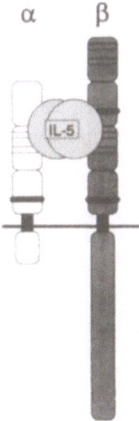

FIG.4., Diagrammic presentation of the IL5/IL5Receptor complex.
The human IL5Rα and -β subunits are presented, emphasizing their
modular structure. Both extracellular and cytoplasmic parts are depicted.
Thick and thin horizontal bars within the FN-III-like modules represent
the WS-WS motif and conserved cysteine residues respectively. The IL5
dimer is also shown.

To localize important hIL5-binding regions on the hIL5Rα subunit, we exploited
the differential cross-species activity of human and mouse IL5. It was shown before that
mouse IL5 binds equally well on the human receptor as on its own, whereas human IL5
binds only very poorly to the mouse receptor. We constructed a panel of chimeric
polypeptides between the human and mouse IL5Rα subunit and expressed these in Cos1
cells. These Cos1 transfectants were then analysed for hIL5 binding. As a control for
structural integrity of the chimeric polypeptides, mIL5 binding was also examined, and
indicated normal binding for all chimaeras. Results are summarized in Figure 5. We
conclude that the first half of the the N-terminal submodule contains the most critical
residues involved in species-specific ligand binding.

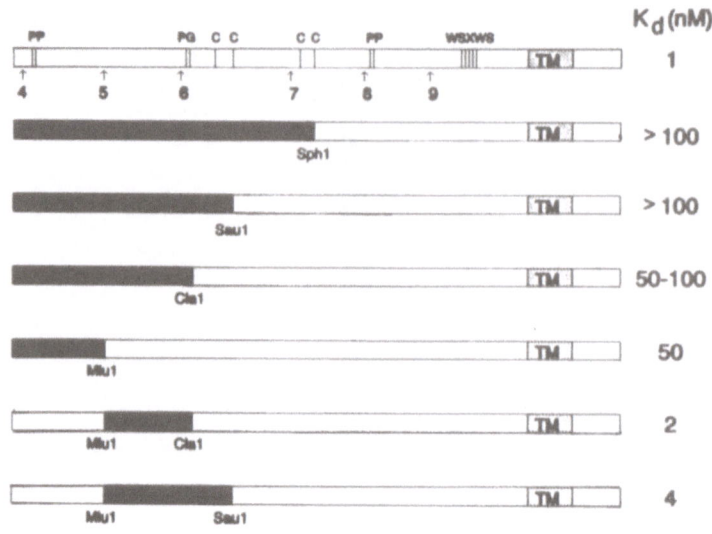

FIG. 5., Human IL5 binding to human/mouse IL5Rα chimaeras.
Bars represent the IL5Rα polypeptides, with open and filled parts
corresponding to human and murine sequences respectively. On the top
bar, conserved, characteristic amino acid residues are indicated in one letter
notation; just below, conserved splice sites are indicated by their number
(see Ref. 24). TM locates the transmembrane anchor. Restriction sites used
to generate the chimaeric polypeptides are also shown. At the right, binding
affinities for human IL5 as determined by Scatchard plot analysis are
presented.

In order to identify individual residues involved in hIL5 binding, all amino acid
residues (fourteen in total) in this part of the hIL5Rα subunit which differ between
human and mouse were converted into their murine counterparts. Again, mutant
receptors were expressed in Cos1 cells and checked for hIL5 binding. Whereas most of
these mutants bound hIL5 with similar affinity as the native receptor, two mutants at
closely linked positions, D56E and E58D, bound hIL5 with significantly lower affinity.
Scatchard plot analyses for these mutant receptors are shown in figure 6.

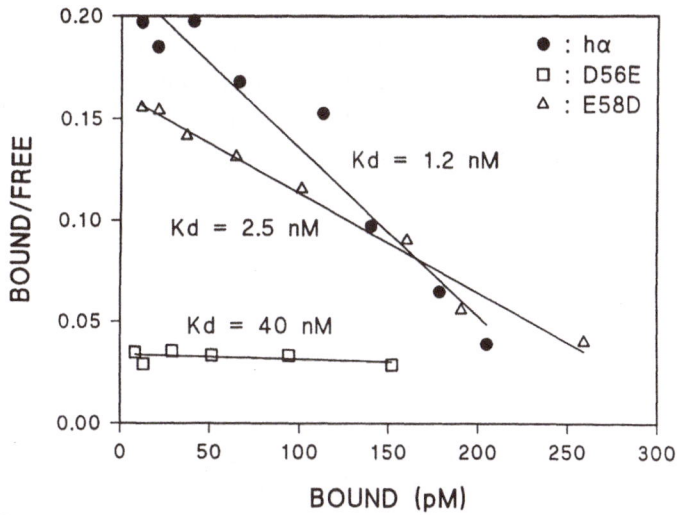

FIG. 6., Scatchard plot analysis of the mutant receptors D56E and E58D. expressed on Cos1 cells.

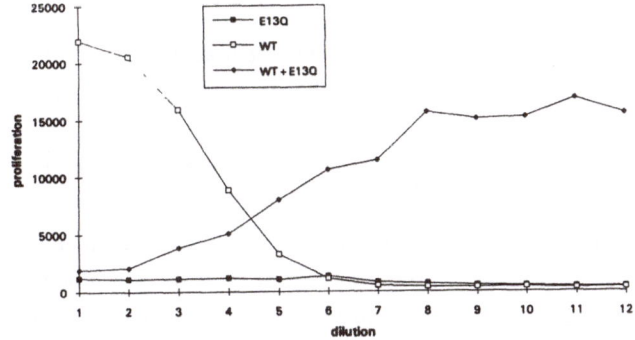

FIG. 7. Antagonistic properties of mutant E13Q on TF-1(hIL5Rα) cells. TF-1(hIL5Rα) cell proliferation was measured by ^{3}H-Thymidine incorporation for serial dilutions of wild-type hIL5 (1:3; open squares, starting concentration is 3.3 ng/ml) and mutant E13Q (1:2; full squares, starting concentration is 500 ng/ml). Diamonds indicate proliferation in the presence of 500 pg wild-type hIL5 in each well, supplemented with a serial dilution of E13Q as before.

ANTAGONISTIC PROPERTIES OF THE hIL5 E13Q MUTEIN

One mutation at position 13, E13Q, was analysed in more detail. This mutant has a specific activity of 10^4 U/mg on the FDCP-1-CA1 cell line, or about 0.1% of the wild type activity. In contrast, on a hIL5-responsive subclone of TF-1 cells (an erythroleukemia cell line originally described by Kitamura et al., ref. 30, which was transfected with the human IL5Rα subunit), this IL5 variant has no detectable proliferative activity. Furthermore, as shown in figure 6, this mutant is a potent antagonist.

CONCLUSIONS

A first requirement for receptor triggering is the binding of hIL5 to the hIL5Rα subunit. By testing a series of site-specific mutations in hIL5 in a solid phase binding assay, residues involved in α-subunit interaction have been identified. These mutants cluster together around the dimer interface belonging either to one or both monomers. Due to the two-fold symmetry of the dimeric IL5 structure, analysis of heterodimers in which different mutations are combined, will be required to find out whether the IL5Rα binding site is formed by one or by both monomers or whether an intermediate situation is present. Irrespective of this uncertainty, we can nevertheless deduce that the area which interacts with the receptor α-subunit, is closely located to the central symmetry axis of the dimer. This interpretation is confirmed by the location of the overlapping epitopes of a non-neutralizing (H30) and a neutralizing (5A5) monoclonal antibody, which bind either apically or to the central axis of the outer β-strand.

Mutations at residue 13 affected the receptor activation without reducing hIL5Rα binding. Moreover, on TF1-hIL5Rα cells, the proliferative activity was abolished and the E13Q mutein behaved as a potent receptor antagonist. The molecular basis for this antagonism is unclear at present. In analogy with the related cytokines IL3 and GM-CSF, where mutations at a homologeous position cause defects in interaction with the β-subunit (26,27,28), one can postulate a similar mechanism for the hIL5 E13Q mutein. Unexpectedly however, we could not observe any differential physicochemical behaviour for this mutein compared to wild type IL5 (including Scatchard plot analysis and chemical cross-linking, data not shown). Alternative explanations include a defect in inducing a critical conformational shift in the receptor complex, or a defect in engaging a third receptor component (which might be a second β-subunit).

The analysis of interspecies chimaeras allowed us also to identify residues on the hIL5Rα-subunit involved in ligand binding. Interestingly, these residues, D56 and E58, are located on the first extracellular cytokine receptor subdomain. This is in apparent contrast to the growth hormone case, the only member of the cytokine family where detailed structural data is available for the ligand/receptor complex (15). In the latter case, the receptor is built up by only two cytokine receptor subdomains and the ligand binds close to the hinge region between them. Full understanding of both similarities and dissimilarities between these two members of the cytokine/cytokine receptor superfamily will await the elucidation of the three-dimensional structure of the IL5/IL5R complex.

ACKNOWLEDGMENTS: We would like to thank the technical assistance of Tania Tuypens, Annick Verhee, Freya Van Houtte and Ina Faché. We are also greatly indebted to Prof. W. Fiers and Prof. C. Sanderson for helpful discussions throughout the course of this work.

REFERENCES

1. Azuma, C., Tanabe, T., Konishi, M., Kinashi, T., Noma, T., Matsuda, F., Yaoita, Y., Takatsu, K., Hammarström, L., Smith, C.I.E., Severinson, E. and Honjo, T. (1986) Nucl. Acids Res., 14, 9149-9158.
2. Tavernier, J., Devos, R., Van der Heyden, J., Hauquier, G., Bauden, R., Faché, I., Kawashima, E., Vandekerckhove, J., Contreras, R. and Fiers, W. (1989) DNA, 8, 491-501.
3 Tsujimoto, M., Adachi, H., Kodama, N., Tsuruoka, N., Yamada, Y., Tanaka, S., Mita, S. and Takatsu, K. (1989) J. Biochem. 106, 23-28.
4. Proudfoot, A.E.I., Fattah, D., Kawashima, E.H., Bernard, A. and Wingfield, P.T. (1990) Biochem. J., 270, 357-361.
5. Minamitake, Y., Kodama, S., Katayama, T., Adachi, H., Tanaka, S. and Tsujimoto, M. (1990) J. Biochem. 107, 292-297.
6. Milburn, M.V., Hassell, A.M., Lambert, M.H., Jordan, S.R., Proudfoot, A.E.I., Graber, P. and Wells, T.N.C. (1993) Nature, 363, 172-176.
7. Guisez, Y., Oefner, C., Winkler, F., Schlaeger, E., Zulauf, M., Van der Heyden, J., PLaetinck, G., Cornelis, S., Tavernier, J., Fiers, W., Devos, R. and d'Arcy, A. (1993) FEBS Letters 331, 49-52.
8. Diederichs, K., Boone, T. and Karplus, P.A (1991) Science, 254, 1779-1782.
9. Walter, M.R., Cook, W.J., Ealick, S.E., Nagabhushan, T.L., Trotta, P.P. and Bugg, C.E. (1992) J. Mol. Biol., 224, 1075-1085.
10. Brandhuber, B.J., Boone, T., Kenney, W.C. and McKay, D.B. (1987) Science, 238, 1707-1709.
11. Bazan, J.F. and McKay, D.B. (1992) Science 258, 1358-1362.
12. Smith, L.J., Redfield, C., Boyd, J., Lawrence, G.M.P., Edwards, R.G., Smith, R.A.G. and Dobson, C.M. (1992) J. Mol. Biol. 224, 899-904.
13. Pandit, J., Bohm, A., Jancarik, J., Halenbeck, R., Koths, K. and Kim, S.-H. (1992), Science 258, 1358-1362.

14. Abdel-Meguid, S.S., Shieih, H.S., Smith, W.W., Dayringer, H.E., Viland, B.N. and Bentle, L.A. (1987) Proc. Natl. Acad. Sci. USA, 84, 6434-6437.

15. de Vos, A.M., Ultsch, M. and Kossiakoff, A.A. (1992) Science, 255, 306-312.

16. Tavernier, J., Devos, R., Cornelis, S., Tuypens, T., Van der Heyden, J., Fiers, W. and Plaetinck, G. (1991) Cell, 66, 1175-1184.

17. Murata, Y., Takaki, S., Migita, M., Kikuchi, Y., Tominaga, A. and Takatsu, K. (1992) J. Exp. Med., 175, 341-351.

18. Takaki, S., Tominaga, A., Hitoshi, Y., Mita, S., Sonoda, E., Yamaguchi, N. and Takatsu, K. (1990) EMBO J., 9, 4367-4374.

19. Devos, R., Plaetinck, G., Van der Heyden, J., Cornelis, S., Vandekerckhove, J., Fiers, W. and Tavernier, J. (1991b) EMBO J., 10, 2133-2137.

20. Takaki, S., Mita, S., Kitamura, T., Yonehara, S., Yamaguchi, N., Tominaga, A., Miyajima, A. and Takatsu, K. (1991) EMBO J., 10, 2833-2838.

21. Kitamura, T., Sato, N., Arai, K.-I. and Miyajima, A. (1991) Cell, 66, 1165-1174.

22. Lopez, A., Eglinton, J.M., Gillis, D., Park, L..S., Clark, S. & Vadas, M.A. (1989) Proc. Natl. Acad. Sci. USA, 86, 7022-7026.

23. Tavernier, J., Tuypens, T., Plaetinck, G., Verhee, A., Fiers, W. and Devos, R. (1992) Proc. Natl. Acad. Sci. USA. 89, 7041-7045.

24. Tuypens, T., Plaetinck, G., Baker, E., Sutherland, G., Brusselle, G., Fiers, W., Devos, R. and Tavernier, J. (1992) Eur. Cytokine Netw., 3, 451-459.

25. Devos, R., Guisez, Y., Cornelis, S., Verhee, A., Van der Heyden, J., Manneberg, M., Lahm, H.-W., Fiers, W., Tavernier, J. and Plaetinck, G. (1993) J. Biol. Chem., 9, 6581-6587.

26. Lopez, A., Shannon, M.F., Hercus, T., Nicola, N., Cambareri, B., Dottore, M., Layton, M.J., Eglinton, L. and Vadas, M.A. (1992) EMBO J., 11, 909-916.

27. Shanafelt, A.B., Miyajima, A., Kitamura, T. and Kastelein, R.A. (1991) EMBO J., 10, 4105-4112.

28. Lopez, A., Shannon, M.F., Barry, S., Phillips, J.A., Cambareri, B., Dottore, M., Simmons, P. and Vadas, M.A. (1992) Proc. Natl. Acad. Sci. USA., 89, 11842-11846.

29. Bazan, J.F. (1990) Proc. Natl. Acad. Sci. USA, 87, 6934-6938.

30. Kitamura, T., Tange, T., Terasawa, T., Chiba, S., Kuwaki, T., Miyagawa, K., Piao, Y., Miyazono, K., Urabe, A. and Takaku, F. (1989) J. Cell. Physiol. 140, 323-334.

AAS 46
Novel Molecular Approaches
to Anti-Inflammatory Theory
© 1995 Birkhäuser Verlag Basel

POTENTIAL PHOSPHOLIPASE A$_2$s INVOLVED IN INFLAMMATORY DISEASES

Edward A. Dennis

Department of Chemistry and Biochemistry
University of California, San Diego, 9500 Gilman Drive, La Jolla, CA 92093-0601

A large variety of membrane receptor mediated events lead to the activation of phospholipase A$_2$ (PLA$_2$) (1) either directly or via secondary mediators as indicated in Figure 1. When PLA$_2$ acts on membrane phospholipids containing an arachidonoyl group at the middle or sn-2 position of the glycerolphosphate backbone, this leads to the production of free arachidonic acid and lysophospholipid (2). The free arachidonic acid serves as a precursor for the cyclooxygenase pathway producing thromboxanes, prostacyclins and other prostaglandins.   Free arachidonic acid also serves as a substrate for the 5-lipooxygenase producing the leukotrienes. The other product, the lysophospholipids, are themselves lytic and must be hydrolyzed further by a lysophospholipase or be reacylated. In some instances, the lysophospholipids are converted to platelet activating factor (PAF), another important lipid mediator.

The action of aspirin and other non-steroid anti-inflammatory drugs (NSAID) on the cyclooxygenase is well known and there is a great deal of current interest in the newly discovered inducible cyclooxygenase, known as COX-II. The pharmaceutical industry has focused considerable efforts on developing inhibitors of the lipoxygenase and receptor antagonists of the leukotrienes and PAF. However, it is clear that if one could inhibit the PLA$_2$, one should, at least in principle, block all three of the inflammatory pathways discussed: the production of prostaglandins, the leukotrienes, and PAF.

Unfortunately, this approach has not been as straightforward as some had hoped because one has to ask: Which phospholipase A$_2$ should be inhibited?

It is now clear that there are many different phospholipase A$_2$s and PLA$_2$ constitutes a diverse family of different enzymes (3). Work in our laboratory and many other laboratories utilizing evolutionary considerations (4) has lead to the classification of Group I, II, and III PLA$_2$s as indicated in Figure 2. All of these enzymes have been isolated as extracellular enzymes. They are secreted, small molecular weight PLA$_2$s. They have an absolute requirement for Ca^{2+} with an apparent binding constant in the

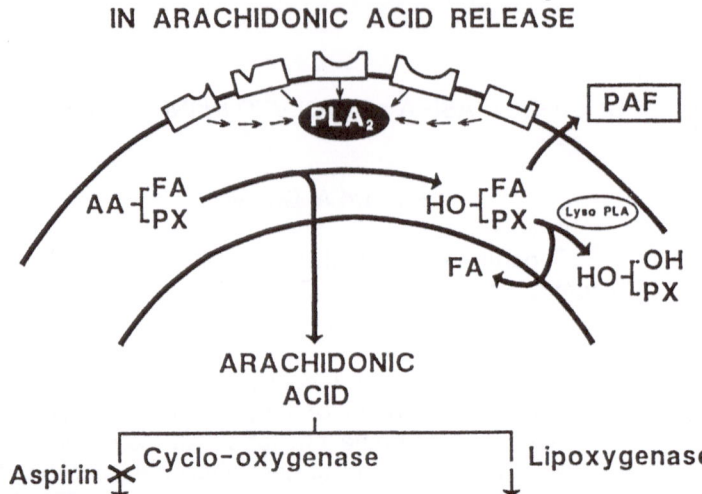

Figure 1: **Central Role of Phospholipase A₂ in Generating Inflammatory Mediators. Adapted from (2).**

GROUP		SOURCE	LOCATION	SIZE	Ca²⁺
I.	A.	Cobras and Kraits	Secreted	13-15 kDa	< mM
	B.	Human Pancreas			
II.	A.	Rattlesnakes and Vipers	Secreted	13-15 kDa	< mM
		Human Synovial/Platelets			
	B.	Gabon Viper			
III.		Bee/Lizard	Secreted	16-18 kDa	< mM
IV.		Raw 264.7/Rat Kidney	Cytosolic	85 kDa	< μM
V.		Canine/Human Myocardium	Cytosolic	40 kDa	None
		Macrophage P388D₁	Cytosolic	80 kDa	None

Figure 2: **Diversity of Phospholipase A₂s. Adapted from (3).**

approximately mM range and a very specific binding site for Ca^{2+} in the catalytic site where Ca^{2+} plays a role in catalysis (5).

In recent years, there has been an increased interest in intracellular PLA$_2$s. The most well studied intracellular PLA$_2$ has been designated as Group IV PLA$_2$ (3). It was originally isolated, sequenced and cloned from the cytosol of human U937 cells independently by Kramer et al. (6) and Clark et al. (7). This enzyme is 85 kDa in size and has a submicromolar requirement for Ca^{2+}. The Ca^{2+} plays a role in association of the PLA$_2$ with its membrane substrate, but in contrast to the secretory PLA$_2$s, is not required for catalysis (8).

The Group IV PLA$_2$ has been referred to as cPLA$_2$ due to its isolation from cytosol. However, it appears to actually act on membranes and it is also found bound to membrane fractions when treated with translocation agents. Other cytosolic PLA$_2$ include Ca^{2+}-independent PLA$_2$s. The most well studied of these have been isolated from canine myocardium by Gross et al. (9) and we (10) have recently described a Ca^{2+}-independent PLA$_2$ from macrophages. Because sequence data on these PLA$_2$s is not available, it would be premature to assign a Group designation.

The characteristics of the macrophage Ca^{2+}-independent PLA$_2$ are summarized in Figure 3. This enzyme is localized in the cytosol (11) and has been purified some 450,000 fold (10). On SDS-PAGE, it appears to be an approximately 80,000 Da protein, but it appears to be active as an oligomer as demonstrated using radiation inactivation techniques and target size analysis. An interesting feature of this enzyme is its activation by ATP and other nucleotides.

It is clear that mammalian cells contain more than one type of PLA$_2$. For example, we (12) have shown that P388D$_1$ macrophages contain a Ca^{2+}-dependent secretory PLA$_2$, probably Group II, a Group IV PLA$_2$, and a Ca^{2+}-independent PLA$_2$. The challenge for the future is to determine the function of these PLA$_2$s in vivo.

- Cytosolic [with LDH in sucrose gradients]
- Purified [450,000 fold]
- SDS Page [M$_r \cong$ 80,000]
- Superose G FPLC [M$_r \cong$ 400,000]
- Substrate [Triton X-100/DPPC mixed micelles 4:1)]
- Specificity [DPPC > PAPC > HAPC]
- Activation [ATP and other nucleotides]

Figure 3: **Characteristics of Ca^{2+} - Independent Phospholipase A$_2$ from Macrophages. Details provided in (10).**

ACKNOWLEDGMENTS
Studies in our laboratory have been supported by grants from NIH GM20,508, GM51,606 and HD26,171.

REFERENCES

1. Dennis, E. A. Phospholipases.(1983) in *The Enzymes, Third Edition, Vol 16* (Boyer, P., ed) pp. 307-353, New York, Academic Press.

2. Dennis, E. A. The Regulation of Eicosanoid Production: Role of Phospholipases and Inhibitors. Bio/Technology 1987; 5: 1294-1300.

3. Dennis, E. A. Diversity of Group Types, Regulation, and Function of Phospholipase A_2 J. Biol. Chem. 1994; 269: 13057-13060.

4. Davidson, F. F. and Dennis, E. A. Evolutionary Relationships and Implications for the Regulation of Phospholipase A_2: from Snake Venom to Human Secreted Forms. J. Mol. Evol. 1990; 31: 228-238.

5. Yu, L. and Dennis, E. A. Critical Role of a Hydrogen Bond in the Interaction of Phospholipase A_2 with Transition-State and Substrate Analogues. Proc. Natl. Acad. Sci. USA 1991; 88: 9325-9329.

6. Sharp, J. D., White, D. L., Chiou, X. G., Goodson, T., Gamboa, G. C., McClure, D., Burgett, S., Hoskins, J., Skatrud, P. L., Sportsman, J. R., Becker, G. W., Kang, L. H., Roberts, E. F., and Kramer, R. M. Molecular cloning and expression of human Ca^{2+}-sensitive cytosolic phospholipase A_2. J. Biol. Chem 1991; 266: 14850-14853.

7. Clark, J. D., Lin, L. L., Kriz, R. W., Ramesha, C. S., Sultzman, L. A., Lin, A. Y., Milona, N., and Knopf, J. L. A novel arachidonic acid-selective cytosolic PLA_2 contains a Ca^{2+}-dependent translocation domain with homology to PKC and GAP. Cell 1991; 65: 1043-1051.

8. Reynolds, L., Hughes, L., Louis, A. I., Kramer, R. A., and Dennis, E. A. Metal Ion and Salt Effects on the Phospholipase A_2, Lysophospholipase, and Transacylase Activities of Human Cytosolic Phospholipase A_2. Biochim. Biophys. Acta. 1993; 1167: 272-280.

9. Hazen, S. L., Stuppy, R. J., and Gross, R. W. Purification and characterization of canine myocardial cytosolic phospholipase A_2. A calcium-independent phospholipase

with absolute f1-2 regiospecificity for diradyl glycerophospholipids. J. Biol. Chem. 1990; 265: 10622-10630.

10. Ackerman, E. J., Kempner, E. S., and Dennis, E. A. Ca^{2+}-Independent Cytosolic Phospholipase A_2 from the Macrophage-Like P388D$_1$ Cells: Isolation and Characterization. J. Biol. Chem. 1994; 269: 9227-9233.

11. Ross, M. I., Deems, R. A., Jesaitis, A. J., and Dennis, E. A. Phospholipase Activities of the P388D$_1$ Macrophage-Like Cell Line. Arch. Biochem. Biophys. 1985; 238: 247-258.

12. Barbour, S. and Dennis, E. A. Antisense Inhibition of Group II Phospholipase A$_2$ Expression Blocks the Production of Prostaglandin E$_2$ by P388D$_1$ Cells. J. Biol. Chem. 1993; 268: 21875-21882.

AAS 46
Novel Molecular Approaches
to Anti-Inflammatory Theory
© 1995 Birkhäuser Verlag Basel

MEDIATION OF INFLAMMATION BY CYCLOOXYGENASE-2

Karen Seibert, Jaime Masferrer, Yan Zhang, Susan Gregory, Gary Olson, Scott Hauser, Kathleen Leahy, William Perkins, and Peter Isakson.

Inflammatory Disease Research
G. D. Searle and Monsanto Corporate Research
800 N. Lindbergh
St. Louis, Missouri, 63167, USA.

Abstract

Non-steroidal antiinflammatory drugs (NSAIDs) are commonly used for the treatment of inflammation, pain, and fever. Mechanistically, these compounds are believed to act via inhibition of the enzyme cyclooxygenase (COX), which catalyzes the conversion of arachidonic acid to the prostaglandins (PGs). Although commercially available NSAIDS are efficacious antiinflammatory agents, significant side effects limit their use. Recently two forms of COX were identified- a constitutively expressed COX-1 and a cytokine-inducible COX-2. Commercially available NSAIDs like indomethacin inhibit both COX-1 and COX-2 suggesting the hypothesis that toxicities associated with NSAID therapy are due to inhibition of the non-regulated or constitutive form of COX (COX-1) in normal tissues, whereas therapeutic benefit derives from inhibition of the inducible enzyme, COX-2, at the site of inflammation. Therefore, a selective inhibitor of COX-2 may be anti-inflammatory without GI toxicity - providing a significant improvement over currently available NSAIDs.

Introduction

NSAIDs are widely used in the treatment of pain and inflammation associated with acute injury or chronic diseases, such as rheumatoid or osteoarthritis. Since the 1970's it has been understood that these compounds as a class inhibit the production of pro-inflammatory prostaglandins through the inhibition of the cyclooxygenase (COX) enzyme(1,2). Once thought to be a single enzyme, at least two forms of COX exist in vivo. COX-1 is constitutively expressed in most tissues, while the inducible form (COX-2) is induced by several growth factors and cytokines, both in vitro and in vivo (3-9). Anti-inflammatory glucocoriticoids, like dexamethasone, selectively inhibit the expression of COX-2 mRNA without an effect on constitutive COX-1 expression (10-12). Unfortunately, although the steroids are potent anti-inflammatory agents, the side-effects of chronic steroid administration significantly limit their therapeutic usefulness. .

The NSAIDs, on the other hand, are widely used to treat the symptoms associated with both acute and chronic inflammatory conditions. There are currently more than twenty five commercially available NSAIDs - including aspirin, indomethacin, and others - all of which exhibit anti-inflammatory and analgesic activity in vivo. Unfortunately, the therapeutic utility of these compounds also is limited by the high incidence of gastric toxicity associated with their use, characterized by erosion of the gastric mucosa, often leading to ulceration and frank hemorrhage (13). This NSAID-induced gastric toxicity is believied to be derived from the inhibition of physiologically important PG production in the gastrointestinal mucosa. With the identification of a novel, cytokine-inducible form of COX, we formulated the hypothesis that constitutive COX-1 activity produces PGs important to normal tissue function in the gut, while inducible COX-2 is the source of pro-inflammatory PGs at the site of tissue injury. Therefore, we anticipate that drugs which selectively inhibit COX-2 will be superior anti-inflammatory agents with clear benefits over existing NSAIDs.

Tissue distribution of COX-1 and COX-2.

Using quantitative mRNA analysis, we examined the relative distribution of COX-1 and COX-2 in both normal and inflamed tissues and report that COX-1 expression dominates normal tissues while COX-2 mRNA is virtually undetectable in normal tissue but induced at the inflammatory site. Specific RNA probes were constructed from the known sequences of murine or rat COX-1 and COX-2, and RNA samples were prepared from numerous tissues (including brain, heart, liver, lung, kidney, spleen, stomach, and colon) for analysis by RNAse protection. COX-1 mRNA was detected in all tissues examined with the highest constitutive levels found in the liver, stomach, spleen, and colon while COX-2 RNA in normal tissues was extremely low in all the tissues examined. We previously reported that in vivo administration of endotoxin (lipopolysaccharide, LPS) in mice resulted in a time-dependent induction of COX expression in peritoneal macrophages coincident with an increase in pro-inflammatory PG in the peritoneal cavity (5). With the availability of specific molecular probes for COX-1 and COX-2 we have now examined the effect of vivo administration of endotoxin (LPS) systemically in the mouse. LPS caused a significant increase in COX-2 mRNA in lung, brain, heart, stomach, spleen, kidney, and colon, with the largest effect observed in the lung; there was not a significant change in COX-1 expression in those tissues examined.

The therapeutic usefulness of the commercial NSAIDs is limited primarily by the toxicity that results from inhibition of PGs in the stomach and gut; therefore, the distribution and function of COX in the GI tract is of particular interest. Probes to detect both rodent and human RNA were engineered and RNAs obtained from mouse, rat, and human GI tissues. Results indicate that the majority of the mRNA expressed in normal gut tissues is COX-1 with slight to undetectable amounts of COX-2 mRNA observed. This observation supports the hypothesis that selective COX-2 inhibitors could potentially spare the physiological production of prostaglandins derived from constitutively expressed COX-1 in the gut and thus reduce the risk of NSAID-induced gastric ulceration.

SC-58125 - a selective COX-2 inhibitor

Development of selective inhibitors of COX-2 that are anti-inflammatory and non-ulcerogenic may represent a significant advance for the treatment of acute and chronic inflammatory disorders. To test this hypothesis, we initiated a program to identify compounds that will selectively inhibit the COX-2 enzyme in vitro without affecting COX-1 activity. The genes that express COX-1 and COX-2 contain several stretches of near identity, with areas of divergence occurring primarily at the amino and carboxy termini of the protein; molecular modeling of the active site based on the recently published x-ray structure of sheep seminal vesicle COX suggests that the active sites of the two enzymes are very similar (14). We cloned and expressed recombinant human COX-1 and COX-2 enzyme and developed an in vitro assay to examine the inhibitory profile of the commercial NSAIDs and novel COX inhibitors. Additionally, we evaluated active inhibitors of COX-2 for anti-inflammatory activity and GI toxicity in vivo. We report the discovery a potent and selective COX-2 inhibitor, SC-58125, (1-[(4-methylsulfonyl)phenyl]-3-trifluoromethyl-5-[(4-fluoro)phenyl]pyrazole) that is anti-inflammatory, analgesic, and devoid of GI toxicity in vivo.

Acute anti-inflammatory and analgesic activity.

Human and rodent recombinant COX-1 and COX-2 enzyme were prepared using a baculovirus expression system, providing a rich source of enzyme for in vitro selectivity profiling of current and novel NSAIDs. Similar to other reports, indomethacin inhibited both COX-1 and COX-2 (IC50, uM: COX-1=.05, COX-2=0.8) consistent with the drug's ability anti-inflammatory and ulcerogenic profile in vivo (Table 1) (15). On the other hand, SC-58125 selectively inhibited COX-2 with no effect on recombinant COX-1 in vitro (IC50, uM: COX-1=.03, COX-2=>100.).

The availability of molecular and pharmacological probes specific for COX-1 and COX-2 provided the tools necessary to examine the distribution of the enzymes in inflamed tissues. We focused on inflammatory models that have been traditionally used to select NSAIDs like edema in the rat carrageenan-induced footpad and adjuvant arthritis, examining whether a selective inhibitor of COX-2 is sufficient to provide anti-inflammatory and analgesic activity in vivo.

Injection of carrageenan into the hindpaw of the rat is the traditional model of evaluating the efficacy of NSAIDs in acute pain and analgesia (16, 17). Following injection of the hindpaw, edema formed rapidly with a maximal effect at about three hours. Extraction of the paw tissue revealed a significant induction of PG production in these paws; mRNA analysis of the paws over the three hour time course showed a substantial increase in COX-2 expression in response to carrageenan with no change in the constitutive COX-1 expressed at the site. Both the mixed COX-1/COX-2 inhibitor, indomethacin, and the selective COX-2 inhibitor, SC-58125, blocked the edema formation and PG production at the site (Table 1). Additionally, both indomethacin and SC-58125 blocked the hyperalgesia in response to stimuli (18). Thus, the inhibition of COX-2 in this acute model of inflammation was sufficient to achieve complete anti-inflammatory and analgesic relief in vivo.

COX-2 is expressed in adjuvant-induced arthritis.

Injection of adjuvant in the footpad of normal, male rats results in a time-dependent cellular infiltration, synovial hyperplasia and edema resembling the clinical manifestations of rheumatoid arthritis (19). Using immunological and molecular tools, we found that COX-2 mRNA and protein were elevated in arthritic rat paws over time without significant changes in COX-1 expression, suggesting that PG production in this model is driven exclusively by COX-2. Selective COX-2 inhibitors (e.g. SC-58125) were as efficacious as mixed COX-1/COX-2 inhibitors (e.g. indomethacin) in therapeutically inhibiting edema in adjuvant-induced model of arthritis as well as blocking the elevated production of PGs in the inflamed paw (Table 1).

Role of COX-1 in GI toxicity

As described earlier, tissue distribution studies indicated that the major form of COX expressed in normal stomach and intestinal samples was COX-1. Our hypothesis is that constitutive COX-1 activity protects the GI tract and inhibition of this activity by current NSAIDs leads to GI toxicity. This hypothesis was examined pharmacologically by comparing the ability of non-selective inhibitors such as indomethacin, and the selective inhibitor SC-58125, to inhibit gastrointestinal PG production and to cause ulcers. For gastric toxicity evaluations fasted rats were dosed orally and the stomach excised five hours later.   SC-58125 did not cause lesions in any rats tested at doses up to 200 mg/kg, while indomethacin caused detectable gastric damage with an ED_{50} of 8 mg/kg (Table 1).

Additionally, we recently reported a close correlation between reduction of stomach PG levels by non-selective COX inhibitors and the appearance of GI lesions; in contrast, selective COX-2 inhibitors had no effect on gastric PG production and did not cause lesions even at doses (200 mg/kg) that greatly exceed the anti-inflammatory dose (<1 mg/kg) (20). Taken together these results provide strong support for the hypothesis that COX-1 is responsible for producing cytoprotective PGs in the GI tract .

Table 1:   Pharmacology of a selective COX-2 inhibitor

	IC50 (uM)			ED50 (mg/kg)		
	COX-1	COX2	Edema	Analgesia	Arthritis	Ulcers
Indo.	.05	0.8	2.0	10.	0.2	8.0
SC-58125	>100	.03	10.	10.	0.4	>200

Summary

There are over twenty five NSAIDs available to treat the pain and discomfort of acute and chronic tissue injury and inflammation. To date, all of these agents produce approximately the same pharmacological efficacy and unfortunately, similar mechanism-based toxicity profiles. A major goal of the pharmaceutical industry is to develop drugs that provide anti-inflammatory and analgesic relief without the untoward effects of these currently available agents. Understanding the role of COX-2 in vivo and the discovery of selective COX-2 inhibitors may provide a major advance in the treatment of these patients with a clear advantage over existing drug therapy.

References

1. Smith, J.B. and A. L. Willis (1971) Aspirin selectively inhibits prostaglandin production in human platelets. Nature [New Biol.] 231: 235-239.

2. Vane, J.R. (1971) Inhibition of prostaglandin synthesis a a mechanism of action for the aspirin-like drugs. Nature [New Biol.] 231: 232-235.

3. Raz, A., A. Wyche, N. Siegel,and P. Needleman, (1988) Regulation of fibroblast cyclooyxgenase synthesis by interleukin-1. J. Biol. Chem. 263, 3022-3028.

4. Fu, J., J.L. Masferrer, K. Seibert, A. Raz, and P. Needleman (1990) The induction and suppression of prostaglandin H2 synthase (Cyclooxygenase) in human monocytes. J. Biol. Chem., 265: 1-4.

5. Masferrer, J.L., B.S. Zweifel, K. Seibert, and P. Needleman. (1992) Selective regulation of cellular cyclooxygenase by dexamethasone and endotoxin in mice. J. Clin. Invest. 86: 1375-1379.

6. Sano, H., T.M. Hla, J.A. Maier, L.J. Crofford, J.P. Case, T. Maciag, and R.L. Wilder (1992), In vivo cycloxoygenase expression in synovial tissues of patients with rheumatoid arthritis and osteoarthritis and rats with adjuvant and streptococcal cell wall arthritis. J. Clin. Invest, 89: 97-108.

7. Xie, W., J.G. Chipman, D.L. Robertson, R.L. Erikson, and D.L. Simmons, (1991) Expression of a mitogen-responsive gene encoding prostaglandin synthase is regulated by mRNA splicing. Proc. Natl. Acad. Sci. U.S.A., 88: 2692-2696.

8. Kujubu, D.A., B.S. Fletcher, C. Varnum, R.W. Lim, and H. Herschman (1991) TIS10, a phorbol ester tumor promoter-inducible mRNA from Swiss 3T3 cells, encodes a novel prostaglandin synthase/cyclooxygenase homologue J. Biol. Chem, 266: 12866-12872.

9. Sirois, J., and J.S. Richards., (1992) Purification and characterization of a novel, distinct isoform of prostaglandin endoperoxide synthase induced by human chorionic gonadotropin in granulosa cells of rat preovulatory follicles. J. Biol. Chem., 267: 6382-6388.

10. Raz, A., A. Wyche, and P. Needleman, (1989) Temporal and pharmacological division of fibroblast cyclooxygnease expression into transcriptional and translational phases. Proc. Natl. Acad. Sci., U.S.A., 86: 1657-1661.

11.   Masferrer, J.L., K. Seibert, B.S. Zweifel, and P. Needleman, (1992) Endogenous glucocorticoids regulate an inducible cyclooxygenase enzyme. Proc. Natl. Acad. Sci., 89: 3917-3921.

12.   Kujubu, D.A., and H. Herschman (1992) Dexamethasone inhibits mitogen induction of the TIS10 prostaglandin synthase/cyclooxygenase gene. J. Biol. Chem., 267: 7991-7994.

13. 24. Allison, M.C., Howatson, A.G., Torrance, C. J., Lee, F.D., and Russell, R., (1992) Gastrointestinal damage associated with the use of non-steroidal antiinflammatory drugs. New Engl. J. Med., 327: 749-754.

14.   Picot, D., Loll, P.J. and Garavito, R.M. (1994)   The X-ray crystal structure of the membrane protein prostaglandin H2 synthase-1. Nature 367: 243-249.

15.   Meade, E.A., Smith, W.L. and DeWitt, D.L. (1993) Differential inhibition of prostaglandin endoperoxide synthase (cyclooxygenase) isozymes by aspirin and other non-steroidal anti-inflammatory drugs. J Biol. Chem. 268: 6610-6614

16. Otterness, I.G. and M.L. Bliven, in Nonsteroidal Anti-inflammatory Drugs, J.G. Lombardino, Ed., (J. Wiley and Sons, New York, 1985).

17.   Winter, C.A., E. A. Risley, and G.W. Nuss, G.W (1962) Proc. Soc. Exp. Biol. Med., 111: 544-552.

18.   Hargreaves, K., Dubner, R., Brown, F., Flores, C., and Joris, J. (1988). A new and sensitive method for measuring thermal nociception in cutaneous hyperalgesia. Pain. 32: 77-88.

19.   Winder, C.V., Lembke, L.A. and Stephens, M.D. (1969) Comparative bioassay of drugs in adjuvant-induced arthritis in rats. Arthritis Rheum. 12: 472-482.

20.   Masferrer, J.L., Zweifel, B., Manning, P.T., Hauser, S.D., Leahy, K.M., Smith, W.G., Isakson, P.C., and Seibert, K. (1994) Selective inhibition of inducible cyclooxygenase-2 in vivo is antiinflammatory and nonulcerogenic. Proc. Natl. Acad. Sci., 91: 3228-3232.

AAS 46
Novel Molecular Approaches
to Anti-Inflammatory Theory
© 1995 Birkhäuser Verlag Basel

Secretory phospholipase A2 inhibitors. Possible new anti-inflammatory agents.

K. Tanaka and H. Arita

Shionogi Research Laboratories, Shionogi & Co., Ltd.,12-4, Sagisu 5-chome, Fukushima-ku, Osaka 553, Japan.

Summary: Secretory phospholipase A2 (sPLA2) is now clearly considered to be involved in the pathogenesis of both experimental and clinical inflammatory processes. This has led academic and pharmaceutical industry researchers to expend enormous efforts to identify specific sPLA2 inhibitors to better understand the role of this enzyme in biological systems and to enable its clinical use in the treatment of inflammation and related disorders. Presented here is a brief review of the biological activity of sPLA2 inhibitors and diseases that may be postulated as their possible targets. Also discussed are problems associated with the evaluation of sPLA2 inhibitors for their selectivity and specificity.

Introduction

Phospholipase A2 (PLA2) is a diverse family of important enzymes which have attracted considerable attention because of their role in the production of potent inflammatory mediators such as prostaglandins, leukotrienes and platelet-activating factor (1, 2). PLA2 enzymes exist in both extracellular and intracellular forms. The best studied extracellular PLA2s, the secretory PLA2 (sPLA2), are 14-kDa calcium-dependent enzymes. sPLA2 can be classified into two types, group I (PLA2-I) and group II (PLA2-II), based on their primary structures. Mammalian PLA2-I abundantly occurs in the pancreas and has long been thought to act as a digestive enzyme (3, 4). The other type, mammalian PLA2-II, has been detected in the synovial fluid of patients with rheumatoid arthritis and has been implicated in a variety of other inflammatory states (5). Mammalian PLA2-II has been purified from human platelets and synovial fluid, and the corresponding human gene was cloned, revealing the existence of only a single gene (6, 7). Overexpression of PLA2-II has been shown to increase arachidonic acid (AA) release from fibroblasts and bulk secretion of the PLA2-II in the cell medium (8). Furthermore, treatment with PLA2-II-specific antisense oligonucleotide decreased PLA2 activity in P388D1 cell homogenates by ~60% and reduced the release of [^{3}H]arachidonic acid and prostaglandin E2 (PGE2) from lipopolysaccharide-stimulated cells to nearly resting cell levels (9). These findings are in line with the hypothesis that extracellular PLA2 at inflamed sites may contribute to the progression of inflammation by the generation of pro-inflammatory lipid mediators. In addition to these findings, intradermal injection of PLA2-II produces an inflammatory reaction in the several animal models. The regulation of PLA2-II activity may therefore achieve important therapeutic use, particularly in inflammatory diseases.

This brief review discusses the relevance of sPLA2 in disease, the biological activities of sPLA2 inhibitors, and problems associated with the assessment of sPLA2 inhibitors with respect to their selectivity and specificity.

Relevance of sPLA2 in disease

The involvement of sPLA2 in the pathophysiology of various diseases has been well documented. Recently, many studies have demonstrated the importance of PLA2-II as a possible regulatory enzyme in the genesis and perpetuation of arthritis (for review, see Ref. 10).

Marked increase in circulating PLA2-II has been documented in experimental endotoxin-shock animals (11). Nakano et al. (12) demonstrated the regulation of PLA2-II gene *in vivo* using endotoxin shock rat. Administration of endotoxin to rats increased both PLA2-II activity in the plasma and levels of PLA2-II mRNA in several tissues including aorta, spleen, lung and thymus. Septic shock in humans is consistently associated with a marked rise in serum sPLA2 activity (13). In all patients, serum sPLA2 levels correlated directly with the magnitude and duration of circulatory collapse. The correlation of serum sPLA2 levels with the increased rise of adult respiratory distress syndrome in patients with sepsis has also been noted (14, 15). sPLA2 may also be critically involved in lung pathophysiology, since snake venom PLA2, given as a bolus injection into either the trachea or jugular vein, induces pronounced morphologic and physiologic changes in guinea pig (16). Snyder et al. (17) demonstrated that human PLA2-II has the ability to induce the release of arachidonic acid and the formation of proinflammatory eicosanoids that contract the pleural strips.

Recently, Andersen et al. (18) reported the detection of transcripts of PLA2-II in human skin and overexpression of PLA2-II in skin from patients with psoriasis. Emmerling et al. (19) reported the involvement of sPLA2 in regulating amyloid precursor protein processing and secretion, which may have important implications for understanding the pathogenesis of Alzheimer's disease. More importantly, the elevation of serum PLA2-II in patients with various malignant tumors was reported, in which incidence and magnitude increased with advancement of the cancer stage (20).

It has been postulated that sPLA2 is associated with the pathology of acute pancreatitis (2), in which the catalytic activity of PLA2 is increased in serum, especially in the severe necrotic form of the disease (21). Nevalainen et al. (22) reported that the catalytic activity of PLA2 showed a highly significant correlation with the concentration of PLA2-II but not with that of PLA2-I. These results suggest that PLA2-I circulates mostly as an inactive precursor form in patients with acute pancreatitis whereas PLA2-II is responsible for the increased catalytic activity of the enzyme and thus might be associated with the pathophysiology of the disease. However, the relative roles of sPLA2 in these diseases cannot be elucidated without specific inhibitors or the appropriate genetic manipulations.

Assessment of sPLA2 inhibitor *in vitro*

The assay of sPLA2 is somewhat complex owing to the water-insoluble nature of the substrate employed (23, 24). The overall characteristics such as packing density and lipid surface charge can have a marked effect on sPLA2 activation and hydrolysis. Radio-labeled substrate sources are derived from two categories: synthetic phospholipids or natural "membrane" lipid forms such as *E. coli* membrane. As synthetic phospholipid aggregates are easily designed and controlled, many sPLA2 inhibitors have been evaluated using synthetic phospholipids as substrate. However, synthetic phospholipids are by no means "natural lipid" presentation forms, hence they usually display poor sensitivity compared with membrane lipids. Moreover, synthetic phospholipid aggregates have very low tolerance to assay manipulations such as pH, ionic strength, substrate concentration or product formation. Radio-labeled *E. coli* membrane has frequently been used for the assessment of sPLA2 inhibitors because of its high sensitivity. Spectrophotometric assays, in general, are desirable because they are rapid, convenient, often continuous, and can be adapted to the assay of large numbers of samples. There have been many attempts to develop a good spectrophotometric assay for sPLA2 using a variety of approaches (24).

Assessment of sPLA2 inhibitor *in vivo*

When a sPLA2 inhibitor has been identified, the next step is to assess its bioactivity. Evaluation of anti-inflammatory activity of sPLA2 inhibitor *in vivo* can be done using routine inflammation animal models. However, while these models can reflect the drug's ability to alter an inflammatory response, they cannot address the specificity of drug action against sPLA2. New models are therefore being developed. Two of the most widely used screening methods are the phorbol ester (TPA) and AA induced rodent ear edema assays that employ topical drug application. Inhibitory activity in the TPA model, but not in the AA ear assay, is often used to argue that a compound is working through specific inhibition of PLA2. In the case of the exogenously applied AA, it is not clear whether it simply acts as a substrate for cyclooxygenase and lipoxygenase (LO) or whether it acts as a stimulant for cell activation and inflammation in a more general manner. The use of sPLA2 administration as the initiating inflammatory insult is the most documented and well characterized of models in terms of evaluating the proinflammatory activity of sPLA2. Another model that has been extensively characterized recently is that using snake venom PLA2 injection into mouse paws (25). The resulting edema formation depends on enzyme concentration and appears to be specific for sPLA2 since the edema could be prevented by enzyme pretreatment with p-bromophenacyl bromide, a nonspecific sPLA2 inhibitor (26). Interestingly, whereas sPLA2 action is the initiating event, the model is also sensitive to antihistamine/antiserotonin agents as well as PAF antagonists and glucocorticoids, suggesting that edema formation is due to a concert of mediator events. Therefore, these studies cannot establish the relevance of PLA2-II as a causative or responsive agents in experimental inflammatory responses.

sPLA2 inhibitors

Contrary to popular belief, many potential inhibitors might be expected to be amphipathic with low critical micelle concentration (27). In this review of sPLA2 inhibitors, we make no attempt to overview the tremendous number of compounds claimed to be inhibitors but will focus on the bound ones with high selectivity or marked activity *in vivo*.

A) Natural products

Natural products, from marine and plant extracts and fermentation broths, have long been a valuable source of chemical leads for drug discovery programs. Manoalide [1], a sesterterpene isolated from the marine sponge (*Lufferiella variabilis*), is a potent inhibitor of sPLA2 from bee venom and human synovial fluid. *In vivo*, manoalide and its synthetic analogues (manoalogue and AGN-190383) have been shown to have a variety of anti-inflammatory actions. However, these compounds have been reported to possess a variety of other activities, such as inhibition of the enzymes 5-LO and phospholipase C (PLC), as well as having effects upon calcium homeostasis. Therefore, the anti-inflammatory action of these compounds is not necessarily due to their sPLA2 inhibitory activity (28). BMS-181162 [2] was reported to inhibit human platelet PLA2 (IC50 = 40 μM) and effectively blocked phorbol ester induced skin inflammation in mice (ED50 = 160 μg/ear). In addition, 2 did not significantly inhibit PI-PLC up to 100 μM (29). Another marine natural product, scalaradial [3], isolated from the sponge (*Cacospongia mollior*), potently inhibited bee venom PLA2 (IC50 = 0.070 μM) in a time-dependent, irreversible manner. 3 also potently inhibited human recombinant PLA2-II (IC50 = 0.070 μM) but displayed weak inhibition of U937 cell 85 kDa-PLA2 (IC50 = 20 μM). *In vivo*, topical application of 3 to mouse ear treated with phorbol ester, not only inhibited edema formation but also the increase in myeloperoxidase activity (an index of cellular infiltration). 3 had little or no effect on AA-induced ear edema or myeloperoxidase which is consistent with action on PLA2 (30).

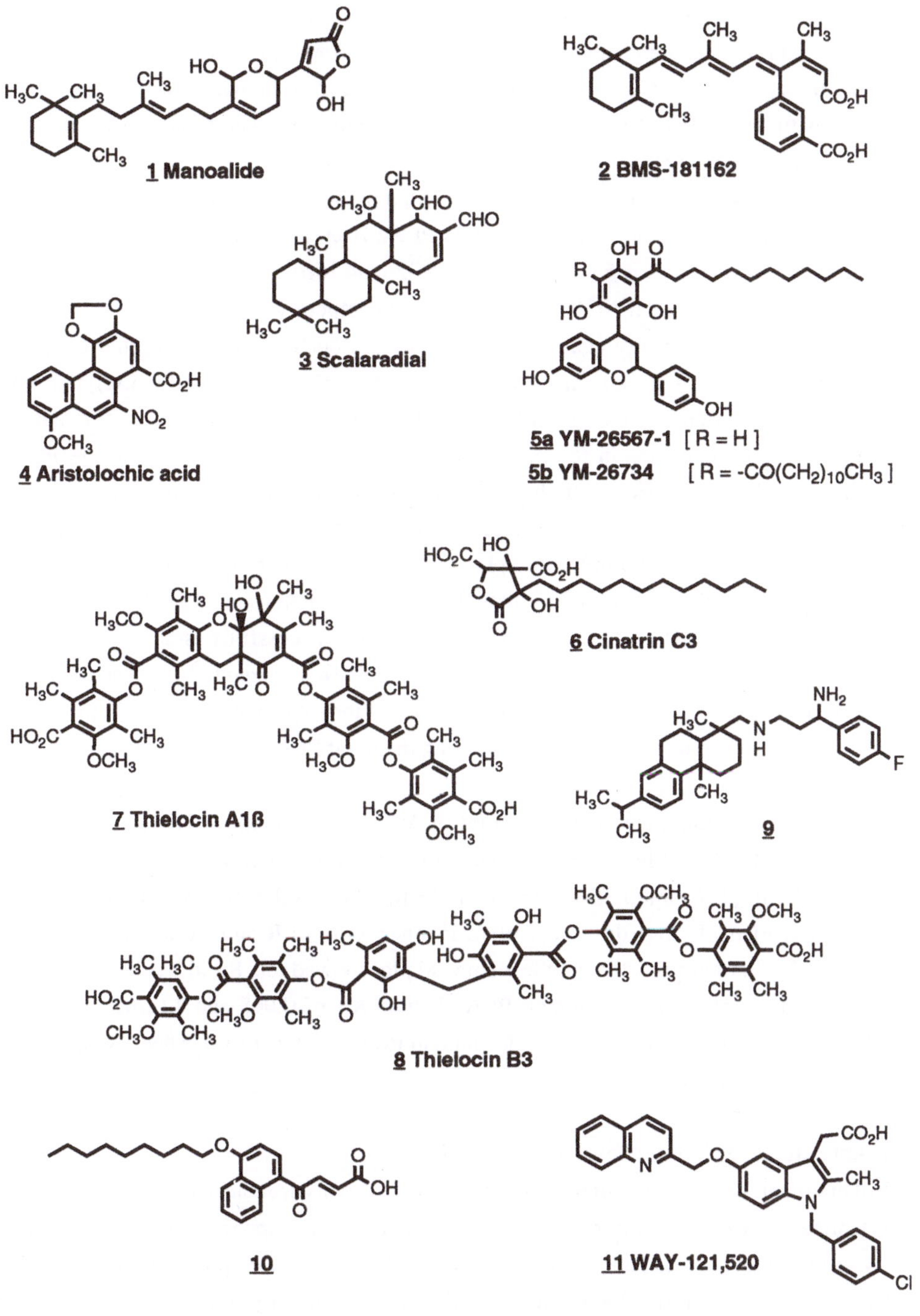

Figure 1. Chemical structures of sPLA2 inhibitors.

In folk medicine, the *Aristolochia* plant and its extracts are used as snake venom antidotes. Aristolochic acid [4], a major chemical component isolated from different species of *Aristolochia* , shows inhibitory activity (IC50 = 85 µM) against PLA2-II from human synovial fluid. When 4 was mixed with human PLA2-II and then injected into mouse foot pad, edema was inhibited in a dose-dependent manner (31).

Miyake et al. (32) reported that YM-26567-1 [5a], a natural product isolated from the fruit of *Horsefieldia amygdaline*, competitively inhibits PLA2-II prepared from rabbit platelets. In the course of further screening for 5a derivatives, YM-26734 [5b] was selected (33). 5b inhibited rabbit platelet-derived PLA2-II (IC50 = 0.085 µM) in a competitive manner (Ki = 0.048 µM) and also showed inhibitory activity against porcine pancreas-derived PLA2-I (IC50 = 6.8 µM). In mice, 5b inhibited TPA-induced ear edema (ED50 = 45 µg/ear and 1 mg/kg, i.v.), but did not decrease AA-induced ear edema (1 mg/ear and 30 mg/kg, i.v.). In rats, the accumulation of exudate fluids and leukocytes in the pleural cavity in response to carrageenan injection was significantly less in a group treated with 5b (20 mg/kg, i.v.) than in the control group. However, it remains to be clarified whether 5b can suppress sPLA2 activity in the inflammatory region.

Some rather weak sPLA2 inhibitors (plastatin, luteospolin, plipastatin, cinatrin C3 [6], folipastatin) have been isolated from the culture broth of microorganisms (34-37).

With the use of PLA2-II purified from rat platelets, we successfully isolated a novel PLA2 inhibitor from the fermentation broth of ascomycetes (*Thielavia erricola* RF-143) and designated it thielocin A1ß [7] (38). Recently, Génisson et al. reported the total synthesis of 7 (39). 7 inhibited various PLA2s in a dose-dependent manner. Among them, rat PLA2-II was the most sensitive to 7 (IC50 = 0.0033 µM). However, 7 showed weak inhibitory activity against rat PLA2-I with an IC50 of 21 µM. In addition, the inhibition of rat PLA2-II was noncompetitive (Ki = 0.0068 µM) and reversible (40). In further screening, we isolated several analogues of 7 from the same fermentation broth (41). Among them, thielocin B3 [8] showed the strongest inhibitory activity toward human PLA2-II in a dose-dependent manner with an IC50 of 0.076 µM (160 times stronger than 7). Furthermore, the double reciprocal plot showed that 8 and 7 behaved kinetically as noncompetitive inhibitors for human PLA2-II with Ki of 0.098 and 12 µM, respectively. Thus, 8 showed 120 times higher affinity for human PLA2-II than 7 (42). These compounds are discussed in more detail below.

B) Agents derived from chemical synthesis

Wilkerson et al. (43) reported a novel series of dehydroabietylamine derivatives as antiinflammatory PLA2 inhibitors. Among the derivatives, Compound [9] showed inhibitory activity against both porcine pancreatic PLA2 (IC50 = 1.4 µM) and rat polymorphonuclear PLA2 (IC50 = 5.6 µM). In addition, 9 exhibited antiinflammatory activity in the rat carrageenan paw edema assay (ED50 = 8.0 mg/kg, p.o.) and in the mouse TPA ear assay (ED50 = 26.5 µg/ear).

However, **9** also showed inhibitory activity against rat basophilic leukemic 5-LO ($IC50 = 4.7$ µM), causing a problem in defining the mechanism of its antiinflammatory activity. **9** was active in the mouse TPA ear assay but inactive in the mouse AA ear test. Therefore, they attributed that the antiinflammatory activity of **9** not to inhibition of 5-LO but of PLA2. Köhler et al. (44) screened 3-(4-alkylbenzoyl)acrylic acid derivatives (ABAAs) in vitro for inhibition of snake venom PLA2. The inhibitory potency of ABAAs increased with the length of the alkyl residues. Compound [**10**] showed the strongest inhibitory activity among the derivatives ($IC50 = 0.07$ µM) with its inhibition being irreversible. In human PMNs, LTB4 and 5-HETE production were essentially reduced by **10** ($IC50 = 35$ and 20 µM, respectively). Since these effects disappeared by addition of AA, the effect of **10** was confirmed to depend on the PLA2 inhibition. Of interest is the fact that immunologically induced bronchospasm in guinea pig was significantly inhibited by **10**. Glaser et al. (45) were intrigued by the report of inhibition of PLA2 by indomethacin, and they synthesized more lipophilic analogs in order to increase the inhibition potency of indomethacin. Among the indomethacin-based inhibitors, WAY-121,520 [**11**] showed the most potent inhibitory activity against human PLA2-II ($IC50 = 4$ µM). Furthermore, *in vivo* antiinflammatory activity was noted in mouse TPA-induced ($ED50 = 91$ µg/ear) and AA-induced (66% inhibition at 400 µg/ear) ear edema. Nevertheless, **11** also shows potent inhibition on 5-LO activity in mouse macrophage (LTC4; $IC50 = 0.004$ µM) and rat PMN (LTB4; $IC50 = 0.010$ µM).

Problems in the assessment of sPLA2 inhibitor

The potential for non-specific and artifactual inhibition of sPLA2 in *in vitro* assay systems should not be underestimated, and this may partly explain why such a wide and confusing variety of structural types have been claimed as being inhibitors. These non-specific sPLA2 inhibitors have been thought to affect the "quality of the interface" by modifying phospholipid bilayer properties that render the phospholipid inaccessible to the enzyme. Davidson et al. (46) reported that lipocortin, which is thought to be an important steroid-inducible inhibitor, inhibits sPLA2 by sequestering the phospholipid substrate; the inhibition can be overcome by high phospholipid substrate concentrations. The same inhibitory mechanism was also indicated by duramycin, a polypeptide antibiotic sPLA2 inhibitor, which specifically interacts with phosphatidyl-ethanolamine (PE). The extent of the inhibition by duramycin was dependent on the substrate concentration (36). Therefore, the sPLA2 inhibitory activity of duramycin is due to direct interaction with the substrate of PE. The inhibitory activities of many sPLA2 inhibitors, including manoalide and acylamino phospholipid analogues, are dependent on the physical state of the substrate (*E. coli* membranes, phospholipids presented as surfactant mixed micelles or sonicated liposomes) and on the type of phospholipid (choline, ethanolamine, serine or inositol). Manoalide reacts with several lysine residues on sPLA2, which reduces the hydrolysis of phosphatidylcholine (PC) but not PE by sPLA2 (47); in fact, PE hydrolysis is enhanced. As mentioned above, duramycin showed inhibitory activity when PE was used as substrate, but not when PI was. When PC was the substrate, it showed a stimulatory effect (Fig. 2a). On the other hand, cinatrin C3 showed inhibitory activity against rat PLA2-II when different phospholipids were used as substrate (Fig. 2b), although its inhibitory activity was affected by the kind of phospholipids. This discrepancy appeared to be due to the amphipathic property of cinatrin C3. Interestingly, thielocin A1ß also showed similar inhibitory activity even when various phospholipids were used as substrate [Fig. 2C]. Therefore, thielocin A1ß may serve as a valuable tool for revealing the substrate-function relationship of sPLA2 (40).

sPLA2 inhibitors are currently assessed using classical inflammatory models, but the crucial question is whether the inhibition of endogenous PLA2 activity at the inflammatory site can result in modulation of an inflammatory event caused by several stimuli. Recently, we demonstrated the antiinflammatory effect of thielocin A1ß and the involvement of PLA2-II in the pathogenesis of rat carrageenan-induced pleurisy (48). When the PLA2-II activity in the exudate (rat pleurisy PLA2-II) was measured using [^{3}H]oleic acid-labeled $E.$ $coli$ membrane, an apparent augmentation was observed 1 h after the injection of carrageenan, with a marked increase until 24 h. At 5 h after coinjection of thielocin A1ß with carrageenan in the pleural cavity, both the number of leukocytes and the amount of protein in the exudate significantly decreased, and the exudate volume in the pleural cavity decreased dose dependently (ED50 = 0.54 mg/kg). Furthermore, rat pleurisy PLA2-II activity also decreased (IC50 = 0.060 mg/kg) by coadministration of thielocin A1ß.

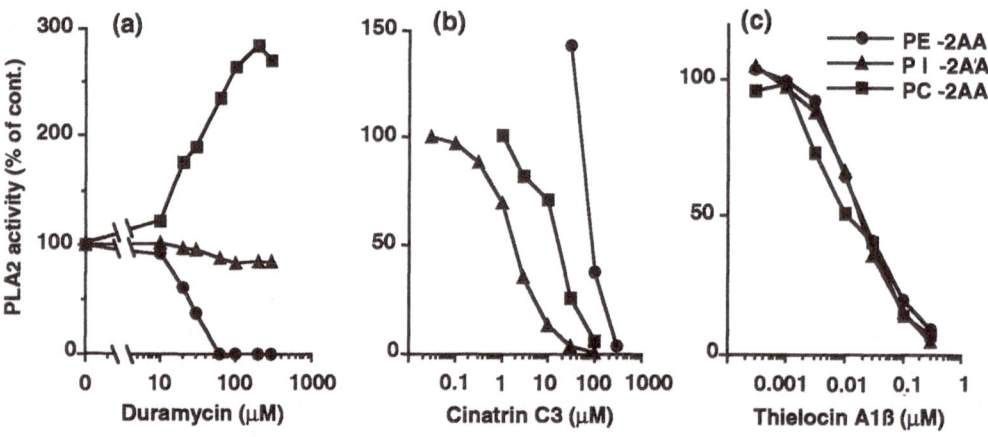

Figure 2. Effect of duramycin, cinatrin C3 and thielocin A1ß on rat group II PLA2 activities toward various phospholipids. Ref. 36, Ref. 40.

The decrease in PLA2-II activity after various doses of thielocin A1ß correlated well with the reduction in the exudate volume (r = 0.85; P < 0.01 ; Fig 3). The exudate volume was also significantly decreased when indomethacin, a cyclooxygenase inhibitor, or dexamethasone, a steroidal antiinflammatory drug, was coinjected with carrageenan. However, these drugs could not significantly attenuate the PLA2-II activity in the exudate, but significantly reduced the levels of PGE2 in the exudate. Of interest, thielocin A1ß had no effect on the PGE2 content in the exudate. These findings suggest that thielocin A1ß possesses antiinflammatory activity, and its mechanism of action may depend on the inhibitory activity of PLA2-II. In addition, the significance of PLA2-II in rat carrageenan-induced pleurisy model was also confirmed by Kakutani et al. (49). Therefore, thielocin A1ß can be considered to be a good tool for studying the implication of PLA2-II in the inflammatory process. Carrageenan-induced pleurisy in rats is an excellent acute inflammatory model in which fluid extravasation, leukocyte migration, prostaglandin synthesis, and PLA2 activity at the inflammatory site can be readily measured in the exudate. However, it should be noted that this model might be far from the clinical situation. Therefore, studies are in progress, with the use of thielocin A1ß and B3, on the role of sPLA2 in the progression of the inflammatory process in animal models that reflect more exactly the clinical situation than carrageenan-induced pleurisy in rats. The crucial issue of clinically relating sPLA2 inhibition to an antiinflammatory effect will have to be addressed.

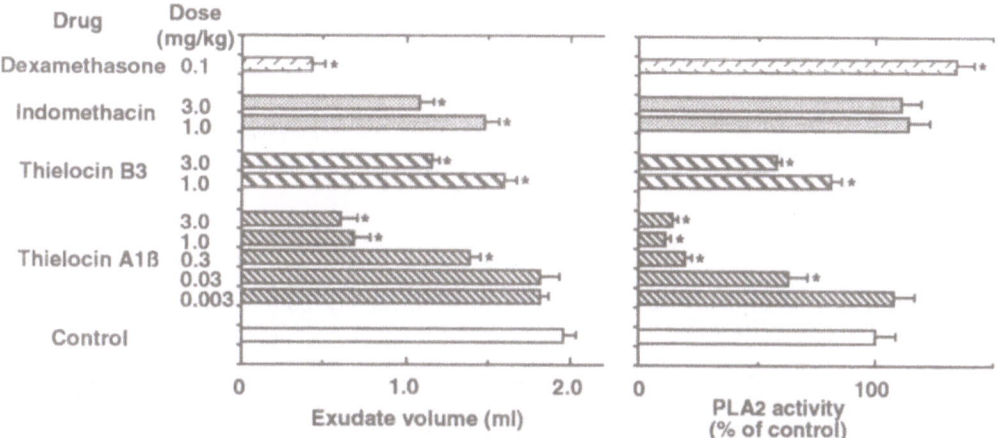

Figure 3. Effects of thielocin A1ß, thielocin B3, indomethacin, and dexamethasone on the volume (left panel) and PLA2 activity (right panel) in the pleural exudate at 5 h after the intrapleural injection of carrageenan to rats. All drugs were administrated with carrageenan. Inhibition of PLA2 activity is expressed as the percentage of the control group. Each value represents the mean ± SEM obtained from six animals. *P < 0.01 compared with respective controls. Ref. 42, Ref. 48.

Conclusion

The roles of sPLA2 in physiological and pathological conditions are important to the targeting of therapeutic intervention in a multitude of diseases involving sPLA2 activity and arachidonic acid release. Certainly, much more work is required, particularly studies performed with selective sPLA2 inhibitors that could be used *in vivo*, to understand the role of this enzyme in the inflammatory responses. The situation is, however, even more complex, since there is an increasing large numbers of structurally different PLA2 enzymes being discovered. Recently, cytosolic PLA2s have been purified from several sources and cDNA encoding a 85 kDa cytosolic PLA2 (cPLA2) has been cloned and expressed (50, 51). Activation of cPLA2 is accompanied by a Ca^{2+}-dependent translocation to membranes in which it may interact with its substrates. cPLA2 preferentially release AA from membrane phospholipids and several agonists, PMA, EGF, TNF and IL-1 increase both AA release in intact cells and PLA2 activity in cell free extracts. These properties of cPLA2 make it a likely candidate for playing a key regulatory role in the arachidonic acid cascade. Moreover, another noticeable PLA2 is the calcium-independent enzyme initially found intracellulary in myocardial tissue. This PLA2, which is also AA selective, has been implicated in ischemic heart disease (52). Since cPLA2 and other intracellular PLA2s have only recently been isolated and characterized, it is not surprising that very few reports of inhibitors against these enzymes have emerged. Future developments will result in the discovery of potent and specific inhibitors of the various PLA2s, that lead to define which enzyme plays in the inflammatory response as the result of regulating the arachidonic acid cascade.

Acknowledgments
We wish to thank Dr. T. Yoshida and all other members of the R1 project, Shionogi Research Laboratories, for their advice and cooperation.

References

1.   Dennis, EA. Phospholipases In: The enzymes. Boyer PD, editor. New York: Academic Press, 1983; 16: 307-353.

2.   Kudo, I., Murakami, M., Hara, S., Inoue, K. Mammalian non-pancreatic phospholipase A2. Biochim Biophys Acta 1993; 117: 217-231.

3.   Nevalainen, TJ. The role of phospholipase A in acute pancreatitis. Scand J Gastroent 1980; 15: 641-650.

4.   Ohara, O., Tamaki, M., Nakamura, E., Tsuruta, Y., Fujii, Y., Shin, M., Teraoka, H, Okamoto, M. Dog and rat pancreatic phospholipase A2: Complete amino acid sequences deduced from complementary DNAs. J Biochem 1986; 99: 733-739.

5.   Vadas, P., Pruzanski, W. Biology of disease. Role of secretory phospholipase A2 in the pathology of disease. Lab Invest 1986; 55: 391-404.

6.   Seilhamer, JJ., Pruzanski, W., Vadas, P., Plant, S., Miller, JA., Kloss, J., Johnson, LK. Cloning and recombinat expression of phospholipase A2 present in rheumatoid arthritic synovial fluid. J Biol Chem 1989; 264: 5335-5338.

7.   Kramer, RM., Hession, C., Johansen, B., Hayes, G., McGray, P., Chow, EP., Tizard, R., Pepinski, RB. Structure and properties of a human non-pancreatic phospholipase A2. J Biol Chem 1989; 264: 5768-5775.

8.   Pernas, P., Masliah, J., Olivier, J.-L., Salvat, C., Rybkine, T., Bereziat, G. Type II phospholipase A2 recombinant overexpression enhances stimulated arachidonic acid release. Biochem Biophys Res Commun 1991; 178: 1298-1305.

9.   Barbour, SE., Dennis, EA. Antisense inhibition of group II phospholipase A2 expression blocks the production of prostaglandin E2 by P388D1 cells. J Biol Chem 1993; 268: 21875-21882.

10. Bomalaski, JS., Clark, MA. Phospholipase A2 and arthritis. Arthritis Rheum 1993; 36: 190-198.

11. Vadas, P., Hay, J. Involvement of circulating phospholipase A2 in the pathogenesis of the hemodynamic changes in endotoxin shock in rabbits. Can J Physiol Pharmacol 1983; 61: 561-566.

12. Nakano, T., Arita, H. Enhanced expression of group II phospholipase A2 gene in the tissues of endotoxin shock rats and its suppression by glucocorticoid. FEBS Lett 1990; 273: 23-26.

13. Vadas, P., Pruzanski, W., Stefanski, E. Extracellular phospholipase A2: Causative agent in circulatory collapse of septic shock? Agents Actions 1988; 24: 320-325.

14. Vadas, P. Elevated plasma phospholipase A2 levels: correlation with hemodynamic and pulmonary changes in gram-negative septic shock. J Lab Clin Med 1984; 104: 873-877.

15. Romaschin, AD., DeMajo, WC., Winton, T., D'costa, Mario., Chang, G., Rubin, B., Gamliel, Z., and Walker, PM. Systemic phospholipase A2 and cachectin levels in adult respiratory distress syndrome and multiple-organ failure. Clin Biochem 1992; 25: 55-60.

16. Durham, SK., Selig, WM. Phospholipase A2-induced pathophysiologic changes in the guinea pig lung. Am J Pathol 1990; 136: 1283-1291.

17. Snyder, DW., Sommers, CD., Bobbitt, JL., Mihelich, ED. Characterization of the contractile effects of human recombinant nonpancreatic secretory phospholipase A2 (PLA2) and other PLA2s on guinea pig lung pleural strips. J Pharmacol Exp Ther 1993; 266: 1147-1155.

18. Andersen, S., Sjursen, W., Lægreid, A., Volden, G., Johansen, B. Elevated expression of human nonpancreatic phospholipase A2 in psoriatic tissue. Inflammation 1994; 18: 1-12.

19. Emmerling, MR., Moore, CJ., Doyle, PD., Carroll, RT., Davis, RE. Phospholipase A2 activation influences the processing and secretion of the amyloid precursor protein. Biochem Biophys Res Commun 1993; 197: 292-297.

20. Ogawa, M., Yamashita, S., Sakamoto, K., Ikei, S. Elevation of serum group II phospholipase A2 in patients with cancers of digestive organs. Res Commun Chem Pathol Pharmacol 1991; 74: 241-244.

21. Büchler, M., Malfertheiner, P., Schädlich, H., Nevalainen, TJ., Friess, H., Beger, HG. Role of phospholipase A2 in human acute pancreatitis. Gastroenterology 1989; 97: 1521-1526

22. Nevalainen, TJ., Grönroos, JM., Kortesuo, PT. Pancreatic and synovial type phospholipase A2 in serum samples from patients with severe acute pancreatitis. Gut 1993; 34: 1133-1136.

23. Van Den Bosch, H. Intracellular phospholipase A. Biochim. Biophys. Acta 1980; 604: 191-246.

24. Reynolds, LJ., Washburn, WN., Deems, RA., Dennis, EA. Assay strategies and methods for phospholipases. Methods Enzymol 1991; 197: 3-23.

25. Marshall, LA., Chang, JY., Calhoun, W., Yu, J., Carlson, RP. Preliminary studies on phospholipase A2 -induced mouse paw edema as amodel to evaluate antiinflammatory agents. J Cell Biochem 1989; 40: 147-155.

26. Kyger, EM., Franson, RC. Nonspecific inhibition of enzymes by p-bromophenacyl bromide. Inhibition of human platelet phospholipase C and modification of sulfhydryl groups. Biocim Biopys Acta 1984; 794: 96-103.

27. Wilkerson, WW. Anti-inflammatory phospholipase A2 inhibitors. Drugs Fut 1990; 15: 139-148.

28. Connolly S., Robinson, DH. A new phospholipase A2 comes to the surface. DN&P 1993; 6: 584-590.

29. Tramposch, KM., Steiner, SA., Stanley, PL., Nettleton, DO., Franson, RC., Lewin, AH., Carroll, FI. Novel inhibitor of phospholipase A2 with topical anti-inflammatory activity. Biochem Biophys Res Commun 1992; 189: 272-279.

30. Marshall, LA., Winkler, JD., Griswold, DE., Bolognese, B., Roshak, A., Sung, C.-M.; Webb, EF., Jacobs, R. Effects of scalaradial, a type II phospholipase A2 inhibitor, on human neutrophil arachidonic acid mobilization and lipid mediator formation. J Pharmacol Exp Ther 1994; 268: 709-717.

31. Vishwanath, BS., Fawzy, AA., Franson, RC. Edema-inducing activity of phospholipase A2 purified from human synovial fluid and inhibition by aristolochic acid. Inflammation 1988; 12: 549-561.

32. Miyake, A., Yamamoto, H., Takebayashi, Y., Imai, H., Honda, K. The novel natural product YM-26567-1: a competitive inhibitor of group II phospholipase A2. J Pharmacol Exp Ther 1992; 263: 1302-1307.

33. Miyake, A., Yamamoto, H., Kubota, E., Hamaguchi, K., Kouda, A., Honda, K., Kawashima, H. Suppression of inflammatory responses to 12 -O-tetradecanoyl-phorbol-13-acetate and carrageenin by YM-26734, a selective inhibitor of extracellular group II phospholipase A2. Br J Pharmacol 1993; 110: 447-453.

34. Singh, PD., Johnson, JH., Aklonis, CA., Bush, K., Fisher, SM., O'Sullivan J. Two new inhibitors of phospholipase A2 produced by *Penicillium chermesinum*. J Antibiot 1985; 38: 706-712.

35. Nishikiori, T., Naganawa, H., Muraoka, Y., Aoyagi, T., Umezawa, H. Plipastatins: new inhibitors of phospholipase A2, produced by *Bacillus cereus* BMG302-fF67. J Antibiot 1986; 39: 755-761.

36. Tanaka, K., Itazaki, H., Yoshida, T. Cinatrins, a novel family of phospholipase A2 inhibitors. II. Biological activities. J Antibiot 1992; 45: 50-55.

37. Hamano, K., Kinoshita-Okami, M., Hemmi, A., Sato, A., Hisamoto, M., Matsuda, K., Yoda, K., Haruyama, H., Hosoya, T., 'Tanzawa, K. Folipastatin, a new depsidone compound from *Aspergillus unguis* as an inhibitor of phospholipase A2. J Antibiot 1992; 45:1195-1201.

38. Yoshida, T., Nakamoto, S., Sakazaki, R., Matsumoto, K., Terui, Y., Sato, H., Arita, H., Matsutani, S., Inoue, K., Kudo, I. Thielocin A1α and A1β, novel phospholipase A2 inhibitors from ascomycetes. J Antibiot 1991; 44: 1467-1470.

39. Génisson, Y., Tyler, PC., Young, RN. Total synthesis of (±)-thielocin A1ß: a novel inhibitor of phospholipase A2. J Am Chem Soc 1994; 116: 759-760.

40. Tanaka, K., Matsutani, S., Matsumoto, K., Yoshida, T. A novel type of phospholipase A2 inhibitor, thielocin A1ß, and mechanism of action. J Antibiot 1992; 45: 1071-1078.

41. Tanaka, K., Matsutani, S., Matsumoto, K., Yoshida, T. Manuscript in preparation.

42. Tanaka, K., Matsutani, S., Kanda, A., Kato, T., Yoshida, T. Thielocin B3, a novel antiinflammatory human group II phospholipase A2 specific inhibitor from ascomycetes. J Antibiot in press.

43. Wilkerson, W., DeLucca, I., Galbraith, W., Gans, K., Harris, R., Jaffee, B., Kerr, J. Antiinflammatory phospholipase A2 inhibitors. I. Eur J Med Chem 1991; 26: 667-676.

44. Köhler, T., Heinisch, M., Kirchner, M., Peinhardt, G., Hirschelmann, R., Nuhn, P. Phospholipase A2 inhibition by alkylbenzoylacrylic acids. Biochem Pharmacol 1992; 44: 805-813.

45. Glaser, KB., Carlson, RP., Sung, A., Bauer, J., Lock, Y.-W., Holloway, D., Sturm, R., Hartman, D., Walter, T., Woeppel, S., Howell, R., Gray, W., Grimes, D., Kubrak, D., Banker, A., Kreft, A., Weichman, BM. Pharmacological characterization of WAY-121,520: a potent anti-inflammatory indomethacin-based inhibitor of 5-lipoxygenase / phospholipase A2. Agents Actions 1993; 39: C30-C35.

46. Davidson, FF., Dennis, EA., Powell, M., Glenney, JR. Jr. Inhibition of phospholipase A2 by "lipocortins" and calpactins. An effect of binding to substrate phospholipids. J Biol Chem 1987; 262: 1698-1705.

47. Dennis, EA. Regulation of eicosanoid production: Role of phospholipases and inhibitors. Biotechnology 1987; 5: 1294-1300.

48. Tanaka, K., Kato, T., Matsumoto, K., Yoshida, T. Antiinflammatory action of thielocin A1ß, a group II phospholipase A2 specific inhibitor, in rat carrageenan-induced pleurisy. Inflammation 1993; 17: 107-119.

49. Kakutani, M., Murakami, M., Okamoto, H., Kudo, I., Inoue, K. Role of type II phospholipase A2 in the inflammatory process of carrageenan-induced pleurisy in rats. FEBS Lett 1994; 339: 76-78.

50. Clark, JD., Lin, L-L., Kriz, RW., Ramesha, CS., Sultzman, LA., Lin, AY., Milona, N., Knopf, JL. A novel arachidonic acid-selective cytosolic PLA2 contains a Ca^{2+}-dependent translocation domain with homology to PKC and GAP. Cell 1991; 65: 1043-1051.

51. Sharp, JD., White, DL., Chiou, XG., Goodson, T., Gamboa, GC., McClure, D., Burgett, S., Hoskins, J., Skatrud, PL., Sportsman, JR., Becker, GW., Kang, LH., Roberts, EF., Kramer, RM. Molecular cloning and expression of human Ca^{2+}-sensitive cytosolic phospholipase A2. J Biol Chem 1991; 266: 14850-14853.

52. Hazen, SL., Stuppy, RJ., Gross, RW. Purification and characterization of canine myocardial cytosolic phospholipase A2. A calcium-independent phospholipase with absolute sn -2 regiospecificity for diradyl glycerophospholipids. J Biol Chem 1990; 265: 10622-10630.

AAS 46
Novel Molecular Approaches.
to Anti-Inflammatory Theory
© 1995 Birkhäuser Verlag Basel

RECENT INSIGHTS INTO THE STRUCTURE, FUNCTION AND BIOLOGY OF cPLA2

Ruth M. Kramer and John D. Sharp

Lilly Research Laboratories, Indianapolis, Indiana, 46285, USA

SUMMARY: The 85-kDa cytosolic PLA2 (cPLA2) is present in most cells and tissues and its structural and functional properties have been described. Different agonists, growth factors and cytokines activate cPLA2 to hydrolyze cellular phospholipids thereby providing the precursor substrates for the biosynthesis of eicosanoids and platelet-activating factor (PAF), the well-known mediators of inflammatory and allergic reactions. Recent studies discussed here suggest that cPLA2 is a receptor-regulated enzyme involved in the inflammatory response. Therefore, inhibitors of cPLA2 may be useful as therapeutic agents in the treatment of inflammatory diseases.

cPLA2 GENE AND PROTEIN

The Ca^{2+}-sensitive cytosolic PLA2 (cPLA2) was first characterized in RAW 264.7 mouse macrophage cells by Leslie and collaborators (1). The cDNA of cPLA2 was subsequently cloned from human and mouse macrophage cell libraries and the promoter of the rat and human cPLA2 genes was recently isolated (2-4). Collectively, the data show that the cPLA2 cDNA comprises a total of 2880 nucleotides, including about 200 nucleotides for the 5'-untranslated region and about 500 nucleotides for the 3'-untranslated region, and encodes a 749-amino acid protein with a predicted molecular mass of 85.2 kDa. The inferred sequence of murine cPLA2 is greater than 95% homologous to the human cPLA2 sequence (3) indicating great structural similarity between cPLA2 from different species. During cDNA cloning and promoter analysis it became apparent that the cPLA2 gene has at least seven introns, one being in the 5'-untranslated region (2-4, and unpublished observations). The locations of the introns within the coding region correspond to amino acid Cys-139, Asp-186, Val-473, Asp-527, Glu-589 and Val-707. Since intron locations often define functional segments of proteins, it is of interest to note that one of the exons encodes fifty-five amino acids centered around Ser-505 (see below). Undoubtedly, more introns will be identified. The rat cPLA2 promoter sequence lacks the well-known TATA and CAAT elements often seen in promoters transcribed by RNA polymerase II that are subject to strong and highly specific expression. In that property it resembles promoters of so-called "housekeeping" genes, named to indicate the fact that these genes are expressed in most cells and are required for basic functions common to all cells. Other features of housekeeping gene promoters are the existence of unmethylated CpG islands, high G+C content in the 5'-flanking region, frequent appearance of GC-rich motifs, such as the Sp1 element GGCGGG, and multiple sites of initiation spread over a large region. While cPLA2 is expressed widely (see below), there are cells that do not express it (5). Although the promoter has a G+C content approaching 50% in the immediate region of initiation of transcription, this level of GC content is not as high as that seen in typical housekeeping genes. Moreover, there are no Sp1 motifs and only one major transcript has been detected so far. Finally, the cPLA2 promoter sequence contains modified recognition sequences for the transcription factors AP-1 and NFκB known to be activated by cytokines, including IL-1 and TNF (see below). Collectively, these observations suggest that the cPLA2 promoter may be classified as a regulated TATA-less promoter, similar to the promoters of certain immunodifferentiation genes (6). The 3'-untranslated region of the cPLA2 mRNA contains regions with increased AU content and multiple conserved AUUUA sequences suggested to regulate mRNA stability. Southern blot analysis of human genomic DNA digests with cDNA fragments as probes and reduced stringency indicated that cPLA2 is present as a single copy gene and that there are no additional genes closely related to cPLA2 (2,3). The human cPLA2 gene is located on chromosome 1 (4).

The deduced cPLA2 protein sequence has several interesting structural features. Firstly, the cPLA2 sequence contains a 109-amino acid segment that exhibits 33% homology with phospholipase B from *P.notatum* (7). Within this domain the sequence Gly-Leu-Ser-Gly-Ser of cPLA2 aligns with the lipase consensus sequence Gly-X(Leu)-Ser-X(Gly)-Gly of phospholipase B. NMR structural studies of a complex of cPLA2 and a trifluoromethyl ketone inhibitor suggested that a serine group may be involved in the catalytic mechanism of cPLA2 (8). Using mutagenesis analysis we have recently demonstrated that the conserved serine within the lipase consensus sequence, Ser-228, is essential for the catalytic activities of cPLA2 (7). Secondly, the cPLA2 sequence comprises a 68-amino acid stretch in the N-terminal portion termed "CaLB domain" that shows 38% sequence homology with the C2 region of protein kinase C (PKC) (2,3). The CaLB domain was also noted in the GTPase activating protein (GAP), phospholipase Cγ_1 and the synaptic vesicle protein p65 (3). This domain was found to be responsible for the Ca^{2+}-dependent binding of cPLA2 to membranes or phospholipid substrate (3). Thirdly, there is a 34-amino acid segment in the middle of the sequence that is deficient in hydrophobic amino acids and may represent a flexible "hinge" region. Fourthly, the cPLA2 contains a proline-rich (>12% of total residues) domain towards the C-terminus that may be responsible for its abnormal electrophoretic mobility on SDS polyacrylamide gels. Fifthly, the protein sequence of cPLA2 contains numerous diverse consensus phosphorylation sites (2) for both serine/threonine and tyrosine protein kinases (2,9). In fact, a regulatory phosphorylation site (Ser-505) has been identified that resides within the MAP kinase recognition sequence Pro-X(Leu)-Ser-Pro (10). The mutated cPLA2 (Ser-505-Ala) was enzymatically active, and its specific activity was 81% and 27% of the control wild-type cPLA2 activity when assayed with phosphatidylcholine-Triton X-100 micelles and sonicated phosphatidylcholine liposomes, respectively. Finally, cPLA2 contains nine free cysteine residues. Since the enzyme is inactivated by preincubation with N-ethyl maleimide (7), it appears that cysteine residues may also be important for the catalytic activity of cPLA2.

BIOCHEMICAL AND CATALYTIC PROPERTIES OF cPLA2

cPLA2 becomes catalytically active in the presence of 0.3 to 2 μM Ca^{2+} as present during cell stimulation and is inactive in the presence of excess chelating agents (9). However, enzymatic activity of cPLA2 can be promoted by high salt in the absence of Ca^{2+} (9). The salt effect is likely due to stimulation of hydrophobic interactions between cPLA2 and phospholipids leading to its association with the substrate in the absence of Ca^{2+}. It was first observed by Leslie et al. (1) that cPLA2 preferentially cleaves arachidonic acid-containing phospholipids and hydrolyzes phosphatidylcholine (PC), phosphatidyl-ethanolamine (PE) and phosphatidylinositol (PI). A detailed analysis of the phospholipid substrate specificity of cPLA2 by Gelb and coworkers comparing different 1-stearoyl-2-arachidonoyl-phospholipids revealed that cPLA2 hydrolyzes with the order of preference PC = PI > PE > phosphatidic acid = phosphatidylserine (11). With PC as substrate the fatty acid preference of cPLA2 was found to be arachidonic (20:4) > linolenic (18:3) > linoleic (18:2) > oleic (18:1) $\geq$ palmitoleic (16:1). The order of preference among PCs containing 20-carbon sn-2 acyl chains was arachidonic (20:4) > homogammalinoleic (20:3) > eicosadienoic (20:2) > eicosenoic (20:1) > eicosanoic (20:0) and there appeared to be a preference for positional isomers with double bonds closest to the sn-2 ester (eicosatrienoic 5,8,11 > 5,8,14 > 5,11,14 > 8,11,14). Interestingly, cPLA2 displays other catalytic activities and was reported to exhibit lysophospholipase and transacylase activities when incubated with LPC micelles (9,12). Incubation of cPLA2 with PC containing dual-labeled fatty acids (1-palmitoyl-2-arachidonoyl-PC), revealed that the lysophospholipase activity on newly formed 1-palmitoyl-lyso-PC was low (9,12).

RECEPTOR ACTIVATION OF cPLA2

Receptor stimulation of many different cell types leads to release of arachidonic acid from cellular phospholipids with subsequent production of eicosanoids and PAF. There is substantial evidence to indicate that cPLA2 represents the intracellular PLA2 that responds to extracellular ligands and liberates arachidonic acid for the conversion to eicosanoids. Thus, Lin and collaborators demonstrated that treatment with either ATP or thrombin of Chinese hamster ovary (CHO) cells overexpressing cPLA2 resulted in increased release of arachidonic acid from cellular phospholipids compared to parental CHO cells demonstrating receptor mediated activation of cPLA2 (13). In contrast, CHO cells overexpressing the secretory PLA2 (sPLA2) failed to show enhanced responsiveness upon receptor activation (11). Several regulatory mechanisms act in concert to promote activation of cPLA2 after receptor stimulation.

Many studies have demonstrated that Ca^{2+} ionophores induce maximal release of arachidonic acid from cellular phospholipids suggesting that receptor-mediated activation of PLA$_2$ is secondary to receptor-mediated activation of phospholipase C (PLC) leading to the formation of inositol phosphates that raise intracellular free [Ca^{2+}] (9). In some cellular systems activation of PLA$_2$ and release of arachidonic acid are highly dependent on the presence of extracellular Ca^{2+} and the influx of external Ca^{2+} via plasma membrane channels appears to be a crucial event in PLA$_2$ activation (9). Purified cPLA$_2$ is activated by submicromolar increments of Ca^{2+} as observed in stimulated cells, but its enzymatic activity is further increased when Ca^{2+} concentrations are raised to millimolar levels (9). Since cPLA$_2$ contains an N-terminal domain that binds to membranes in a Ca^{2+}-dependent fashion (3), increased cytosolic free Ca^{2+} mobilized from intracellular stores and/or derived from increased influx of extracellular Ca^{2+} is likely to induce the association of cPLA$_2$ with cellular membranes. In fact, physiological stimuli, such as thrombin and INF-γ cause such redistribution of cPLA$_2$ from the cytosol to the membrane fraction in human platelets (9) and human bronchial epithelial cells (14), respectively.

Phosphorylation also plays an important role in the regulation of cPLA$_2$ activity. Several observations support the notion that PKC may be an important component in this process. Involvement of PKC in the regulation of PLA$_2$ has been demonstrated in mesangial cells, neutrophils, platelets, macrophages, vascular endothelial cells and Madin-Darby canine kidney cells. Furthermore, inhibition of expression of PKC$_\alpha$ by antisense cDNA was found to decrease PLA$_2$-mediated release of arachidonic acid in response to phorboldiester (9). The PLA$_2$ activity in cytosolic extracts from activated cells is increased several-fold compared to that extracted from unstimulated control cells suggesting a stable modification of the enzyme (9,13). Furthermore, a variety of agents, including ATP, thrombin, phorboldiester and PDGF stimulated the phosphorylation of cPLA$_2$ on serine residues in different cells types (13). In platelets, thrombin induced a time- and dose-dependent phosphorylation of cPLA$_2$ that resulted in enhanced catalytic activity, as well as a change in the electrophoretic and chromatographic properties of cPLA$_2$ (15). By comparing the functional properties of cPLA$_2$ from control and thrombin-stimulated platelets, it was found that while phosphorylated cPLA$_2$ exhibited the same Ca^{2+}-requirement and apparent substrate affinity (K$_m$), its catalytic activity (V$_{max}$) was increased compared to control cPLA$_2$. In renal mesangial cells EGF increased PLA$_2$ activity in cell-free extracts, despite the fact neither PI-specific

PLC was activated or cytosolic free [Ca^{2+}] was increased (9). Down-regulation of PKC did not affect the EGF-induced enhancement of PLA2 activity further suggesting that PLA2 may also be stimulated by PKC-independent mechanisms (9). In mouse HEL-30 keratinocytes TGFα promoted an increase in cPLA2 activity that was independent of PKC activation (16). By comparing cells transfected with normal EGF receptor vs. defective EGF receptor, it was found that intrinsic tyrosine kinase activity is required for the PKC-independent activation of PLA2 (9). However, while EGF was unable to promote PLA2 activation in cells containing a defective EGF receptor, Ca^{2+} ionophore was still able to do so (9). It thus appeared that cPLA2 may be a physiological substrate for different types of kinases or, alternatively, for a kinase that is activated by both PKC dependent and independent mechanisms. One such protein kinase is the mitogen-activated kinase (MAP kinase). Lin and coworkers (10) showed that CHO cells transfected with cPLA2 lacking the MAP kinase phosphorylation site (Ser-505-Ala) exhibited diminished ability to release arachidonic acid in response to agonists compared to CHO cells expressing comparable levels of unmutated cPLA2. Unexpectedly, arachidonic acid mobilization by cells overexpressing the mutated cPLA2 (Ser-505-Ala) was also attenuated upon treatment with Ca^{2+} ionophore, despite the fact that this cPLA2 mutant showed enzymatic activity *in vitro*. Hence, in the CHO cell system the phosphorylation of cPLA2 by MAP kinase may be a prerequisite for its activation, and cytosolic levels of free Ca^{2+} may play a secondary role in the regulation of cPLA2. It remains to be established whether MAP kinase is also involved in the activation of endogenous cPLA2 in primary cells, such as macrophages, neutrophils and platelets.

G proteins may be involved in the control of cPLA2 independent of PLC- and/or PKC-dependent pathways (17). Johnson and coworkers (18) reported recently that a functional G_{i2} protein is necessary for full activation of cPLA2 and this activation appears to be independent of the regulation by elevation of intracellular Ca^{2+} and phosphorylation by MAP kinase.

TISSUE DISTRIBUTION AND SUBCELLULAR LOCALIZATION OF cPLA2

Previous work on the purification and characterization of cPLA2 has indicated that monocyte/macrophage cell lines, as well as kidney and spleen contain cPLA2 (9). The cPLA2 gene is expressed widely as quantitated by RT-PCR analysis (5). The levels of cPLA2 mRNA were more prominent in brain, lung, kidney, heart and spleen than in liver. Furthermore, using anti-cPLA2 antibodies the presence of cPLA2 protein was demonstrated in a variety of tissues from the guinea pig, in particular lung, spleen, brain and kidney. The amounts of cPLA2 detected in guinea pig heart and liver were significantly smaller. Recently, cPLA2 was localized to astrocytes in the gray matter in human brain (9). Many different cell types contain cPLA2, namely fibroblasts, platelets, neutrophils, monocytes, alveolar epithelial cells, renal mesangial cells and keratinocytes (9). Using anti-human cPLA2 antibodies cPLA2 can be readily detected in platelets of different species, including dog, guinea pig, rabbit, rat, mouse and sheep (unpublished observation), thus providing further evidence for the structural similarity of cPLA2 in different species. Analysis of the subcellular distribution of cPLA2 in resting and stimulated cells, including platelets and alveolar epithelial cells indicated that upon cell activation the cPLA2 redistributes from the cytosolic to the membrane fraction (9,14). In rat peritoneal macrophages stimulation with Ca^{2+} ionophore resulted in a redistribution of cPLA2 from the cytosolic fraction to nuclear membranes (19).

cPLA2 AND DISEASE

Eicosanoids and PAF are involved in the pathophysiology of inflammation, allergy and other disorders. Specific agonist-receptor or antigen-antibody interactions induce the activation of cPLA2 to hydrolyze cellular phospholipids thereby liberating the precursors for the biosynthesis of these lipid mediators. Since cPLA2 is an early critical enzyme in the cascade of events that lead to the generation of eicosanoids and PAF, its regulation by cytokines and other pro-inflammatory agents has become a subject of great interest.

IL-1 and TNF are pro-inflammatory cytokines that are potent stimuli of eicosanoid and PAF production. Lin and coworkers (20) reported that treatment of human lung fibroblasts with IL-1 resulted in a large increase in the cellular content of cPLA2 (without affecting cyclo-oxygenase levels) that correlated closely with increased cellular PGE2 production. As demonstrated in time-course experiments, treatment of lung fibroblasts with IL-1 for 30 min first stimulated phosphorylation of cPLA2 and then, after 8 to 24 hours, continuously increased protein levels of cPLA2. Dexamethasone inhibited both the induction of cPLA2 biosynthesis and the production of PGE2 by IL-1 in a dose-dependent manner, but did not affect the basal levels of cPLA2. IL-1 is thought to be important in the pathogenesis of glomerular inflammation and injury that is accompanied by enhanced prostaglandin synthesis. Mesangial cells are an abundant source of prostaglandins and, in culture, express a phenotype that mimics their characteristics within the inflamed glomerulus. Schalkwjik et al. (21) and Gronich and coworkers (22) reported that IL-1 activates cPLA2 in mesangial cells by inducing phosphorylation and increasing its intrinsic catalytic activity. Furthermore, upon prolonged incubation (24 hours) IL-1 increased the mRNA and protein level of cPLA2. The effect of IL-1 on cPLA2 activity in mesangial cells was suppressed in the presence of dexamethasone. Hence, activation of cPLA2 may be a key step in IL-1-stimulated synthesis of lipid mediators in mesangial cells. Studies on the effect of TNF on cPLA2 were performed with Hela cells and demonstrated that TNF, like IL-1, induces rapid (within minutes) phosphorylation of cPLA2 that is followed within hours by elevation of cPLA2 protein mRNA and protein (23). The second phase of TNF-induced activation of cPLA2 was suppressed by dexamethasone.

M-CSF (macrophage colony stimulating factor) activated monocytes are thought to play a role in the inflammatory response. M-CSF is required for the growth and differentiation of monocytes and stimulates arachidonic acid mobilization and prostaglandin production in monocytes (24). M-CSF promoted phosphorylation of cPLA2 and increased the levels of cPLA2 mRNA and protein in monocytes (24). The effect of M-CSF on arachidonic acid mobilization and gene expression was biphasic with rapid (30 min) and late (48 hours) stimulation. The M-CSF induced cPLA2 gene expression was, at least in part, due to post-transcriptional stabilization of cPLA2 mRNA.

Bacterial lipopolysaccharide (LPS) provokes a range of immediate and delayed responses in neutrophils and primes this cell for enhanced liberation of arachidonic acid and LTB4 production in response to other stimulating agents, such as opsonized zymosan. LPS priming is accompanied by phosphorylation of cPLA2 and a two to three-fold increase in its intrinsic activity (25). The extent and kinetics of the effect of LPS on cPLA2 parallels the priming of arachidonic acid mobilization suggesting that in neutrophils cPLA2 provides arachidonic acid for the generation of the potent inflammatory mediator LTB4.

Interferons (IFNs) are a heterogenous family of cytokines with multiple activities exerted primarily by inducing the synthesis of proteins. Shelhamer and colleagues (14) reported that IFN-γ, a cytokine mainly released by T-lymphocytes, induced the synthesis and prolonged activation of cPLA2 in human bronchial epithelial cells. Thus, cPLA2 is a protein induced by the IFN family and may play a critical role in airway epithelium currently thought to be involved in initiating and modulating airway inflammation in disorders such as asthma.

Taken together, the above studies demonstrate that the increase in cellular eicosanoid production promoted by cytokines and other pro-inflammatory agents is, at least in part, due to activation of cPLA2 and elevation of its cellular levels. These findings suggest that cPLA2 participates in the inflammatory response. Furthermore, it has become clear that cPLA2 plays an important role both in the rapid and prolonged cellular responses occurring during inflammatory processes. Finally, it appears that cPLA2 may be a target of anti-inflammatory glucocorticoids well-known for their ability to attenuate eicosanoid synthesis in a number of different cell types.

SUMMARY AND CONCLUSION

The Ca^{2+}-sensitive 85-kDa cytosolic PLA2 (cPLA2) is a receptor-regulated enzyme that initiates the cascade of events leading to the production of eicosanoids and PAF. At least two early receptor-mediated events determine the full activation of cPLA2: (1) increased cytosolic free [Ca^{2+}] promoting association of cPLA2 with its membrane phospholipid substrate and (2) phosphorylation by receptor-activated kinases converting cPLA2 into an enzyme of enhanced catalytic efficiency. Arachidonic acid release and eicosanoid production are hallmarks of inflammation, and increased eicosanoids mediate both pathophysiological alterations and cellular processes which lead to inflammatory injury. There is substantial evidence to indicate that cPLA2 may be an important component in the cascade of events leading to the production of eicosanoids during acute and chronic inflammation. Thus, bacterial lipopolysaccharide and pro-inflammatory cytokines potentiate the mobilization of arachidonic acid and subsequent eicosanoid production by inducing post-translational modification(s) and *de novo* synthesis of cPLA2. Hence, cPLA2 may serve a key role in the generation of eicosanoids and PAF in inflammatory and other disease states. Accordingly, cPLA2 is an attractive target for the development of novel anti-inflammatory therapies.

ACKNOWLEDGMENTS

Due to space and reference limitations we cited recent reviews where possible and apologize to those authors whose reports could not be listed individually. We thank Joe Jakubowski for reviewing the manuscript.

REFERENCES

1.   Leslie CC, Voelker DR, Channon JY, Wall MM, Zelarney PT. Properties and purification of an arachidonoyl-hydrolyzing phospholipase A2 from a macrophage cell line, RAW 264.7. Biochim.Biophys.Acta 1988; 963:476-492

2.   Sharp JD, White DL, Chiou XG, Goodson T, Gamboa GC, McClure D, Burgett S, Hoskins J, Skatrud PL, Sportsman JR, Becker GW, Kang LH, Roberts EF, Kramer RM. Molecular cloning and expression of human Ca^{2+}-sensitive cytosolic phospholipase A2. J.Biol.Chem. 1991; 266: 14850-14853

3.   Clark JD, Lin L-L, Kriz RW, Ramesha CS, Sultzman LA, Lin AY, Milona N, Knopf JL. A novel arachidonic acid-selective cytosolic PLA2 contains a Ca^{2+}-dependent translocation domain with homology to PKC and GAP. Cell 1991; 65: 1043-1051

4.   Tay A, Maxwell P, Li Z, Goldberg H, Skorecki K. Isolation of promoter for cytosolic phospholipase A2 (cPLA2). Biochim.Biophys.Acta. 1994; 1217: 345-347

5.   Sharp JD, White DL. Cytosolic PLA2: mRNA levels and potential for transcriptional regulation. J.Lipid Med. 1993; 8: 183-189

6.   Smale ST, Baltimore D. The "initiator" as a transcription control element. Cell 1989; 57: 103-113

7.   Sharp JD, Pickard RT, Chiou XG, Manetta JV, Kovacevic S, Miller JR, Varshavsky AD, Roberts EF, Strifler BA, Brems DN, Kramer RM. Serine-228 is essential for catalytic activities of cPLA2. J.Biol.Chem: in press

8.   Trimble LA, Street IP, Perrier H, Tremblay NM, Weech PK, Bernstein MA. NMR structural studies of the tight complex between a trifluoromethyl ketone inhibitor and the 85-kDa human phospholipase A2. Biochem. 1993; 32: 12560-12565

9.   Kramer RM. Structure and function of cytosolic phospholipase A2. In: Signal-activated phospholipases. Liscovitch M, editor. Austin: R.G. Landes Biomedical Publishers Company, 1994:13-30.

10. Lin L-L, Wartmann M, Lin AY, Knopf JL, Seth A, Davis RJ. cPLA2 is phosphorylated and activated by MAP kinase. Cell 1993; 72: 269-278

11. Hanel AM, Schuttel S, Gelb MH. Processive interfacial catalysis by mammalian 85-kDa phospholipase A2 enzymes on product-containing vesicles: application to the determination of substrate preferences. Biochem. 1993; 32: 5949-5958

12. Leslie, CC. Kinetic Properties of a high molecular mass arachidonoyl-hydrolyzing phospholipase A2 that exhibits lysophospholipase activity. J.Biol.Chem. 1991; 266: 11366-11371

13. Lin L-L, Lin AY, Knopf JL. Cytosolic phospholipase A2 is coupled to hormonally regulated release of arachidonic acid. Proc.Natl.Acad.Sci 1992; 89: 6147-6151

14. Wu T, Levine SJ, Laurence MG, Logu C, Angus C.W, Shelhamer JH. Interferon-γ induces the synthesis and activation of cytosolic phospholipase A2. J.Clin.Invest. 1994; 93: 571-577

15. Kramer RM, Roberts EF, Manetta JV, Hyslop PA, Jakubowski JA. Thrombin-induced phosphorylation and activation of Ca^{2+}-sensitive cytosolic phospholipase A2 in human platelets. J.Biol.Chem. 1993; 268: 26796-26804

16. Kast R, Furstenberger G, Marks F. Activation of cytosolic phospholipase A2 by transforming growth factor-α in HEL-30 keratinocytes. J.Biol.Chem. 1993; 268: 16795-16802

17. Exton JH. Phosphatidylcholine breakdown and signal transduction. Biochim.Biophys.Acta 1994; 1212: 26-42

18. Winitz S, Gupta SK, Qian NX, Heasley LE, Nemenoff RA, Johnson GL. Expression of a mutant G_{12} subunit inhibits ATP and thrombin stimulation of cytoplasmic phospholipase A_2-mediated arachidonic acid release independent of Ca^{2+} and mitogen-activated protein kinase regulation. J.Biol.Chem 1994; 269: 1889-1895.

19. Peters-Golden M, McNish RW. Redistribution of 5-lipoxygenase and cytosolic phospholipase A_2 to the nuclear fraction upon macrophage activation. Biochem.Biophys.Res.Commun. 1993; 196: 147-153

20. Lin L-L, Lin AY, DeWitt DL. Interleukin-1α induces the accumulation of cytosolic phospholipase A_2 and the release of prostaglandin E_2 in human fibroblasts. J.Biol.Chem. 1992; 267: 23451-23454

21. Schalkwijk C, Vervoordeldonk M, Pfeilschifter J, van den Bosch H. Interleukin-1β-induced cytosolic phospholipase A_2 activity and protein synthesis is blocked by dexamethasone in rat mesangial cells. FEBS Letters 1992; 333: 339-343

22. Gronich J, Koieczkowski M, Gelb MH, Nemenoff RA, Sedor JR. Interleukin 1α causes rapid activation of cytosolic phospholipase A_2 by phosphorylation in rat mesangial cells. J.Clin.Invest. 1994; 93: 1224-1233

23. Hoeck WG, Ramesha CS, Chang DJ, Fan N, Heller RA. Cytoplasmic phospholipase A_2 activity and gene expression are stimulated by tumor necrosis factor: Dexamethasone blocks the induced synthesis. Proc.Natl.Acad.Sci. 1993; 90: 4475-4479

24. Nakamura, T, Lin, L-L, Kharbanda, S, Knopf, JL, Kufe, D. Macrophage colony stimulating factor activates phosphatidylcholine hydrolysis by cytoplasmic phospholipase A_2. EMBO Journal 1992; 11: 4917-4922

25. Doerfler ME, Weiss J, Clark JD, Elsbach P. Bacterial lipopolysaccharide primes human neutrophils for enhanced release of arachidonic acid and causes phosphorylation of an 85-kD cytosolic phospholipase A_2. J.Clin.Invest. 1994; 93, 1538-1591

AAS 46
Novel Molecular Approaches
to Anti-Inflammatory Theory
© 1995 Birkhäuser Verlag Basel

CONTROL OF INFLAMMATORY PROCESSES
BY CELL-IMPERMEABLE INHIBITORS OF PHOSPHOLIPASE A$_2$.

Saul Yedgar[1], Phyllis Dan[1], Arie Dagan[1], Isaac Ginsburg[2], Izidore S. Lossos[3]
and Raphael Breuer[3].

[1]Department of Biochemistry, Hebrew University-Hadassah Medical School,
[2]Department of Oral Biology, Hebrew University-Hadassah School of Dental Medicine, and
[3]Pulmonary Research Laboratory, Hadassah University Hospital, Jerusalem, Israel 91120.

ABSTRACT

Cell-impermeable inhibitors of phospholipase A$_2$ were prepared by linking inhibiting molecules to macromolecular carriers which prevent the inhibitor's internalization. These preparations inhibit the release of oxygen reactive species from neutrophils and cell death induced by inflammatory agents, as well as bleomycin-induced lung injury.

INTRODUCTION

The inflammatory process involves the release of inflammatory agents, such as lysins from bacteria, and oxygen reactive species (ORS - hydrogen peroxide, free radicals) and hydrolytic enzymes from activated white cells. These agents cooperatively attack host cells and induce cell and tissue damage (1).

Phospholipase A$_2$ (PLA$_2$) plays a key role in different steps of the inflammatory process (2,3,4). The endogenous enzyme, as the enzyme responsible for the release of arachidonic acid from cell membrane phospholipids, plays a regulatory role in the production of eicosanoids involved in cell activation. It is well established today that the various eicosanoids are involved in development of many pathological processes, mainly inflammatory and allergic processes and cardiovascular disorders (2,3). In addition, in various pathological states, such as arthritis, sepsis and pancreatitis, excessive amounts of

PLA$_2$ in the extracellular fluid and plasma may exogenously hydrolyse cell membrane phospholipids and induce massive cell lysis and tissue damage (2,5).

Accordingly, it has been long proposed to employ PLA$_2$ inhibitors for control of inflammatory diseases. A number of inhibitors have been proposed for regulation of PLA$_2$ activity at the cell membrane. However these inhibitors enter the cell and thus disturb the phospholipid metabolism which is vital for cell viability. For adequate treatment it is necessary to apply a cell-impermeable PLA$_2$ inhibitor which affects the enzyme activity at the cell membrane but does not enter the cell (6).

To fulfill these requirements we have designed and synthesized cell-impermeable inhibitors of PLA$_2$ (PLIs). This was obtained by linking known and novel PLA$_2$ inhibiting molecules to macromolecular (polymeric) carriers, in a way which enables the interaction of the inhibiting moiety with the cell membrane, but its internalization is prevented by the impermeable carrier. As previously described (7,8) a number of PLIs were prepared according to this concept, and they were found efficient in inhibiting the hydrolysis of cell membrane phospholipids and eicosanoid production in different cell systems (8,9). The present paper presents the effect of the cell-impermeable inhibitor CME, composed of N-derivatized phosphatidyl ethanolamine (PE) linked to carboxymethyl cellulose (CMC) (8), on the release of ORS from neutrophils and on the action of inflammatory agents on host cells and tissues, in culture cells and intact animals.

EXPERIMENTAL

Phosphatidyl ethanolamine linked to carboxymethylcellulose (CME) was synthesized as previously described (8).

Cell culture: Human fibroblasts, a generous gift from Prof. G. Bach of the Department of Human Genetics, the Hadassah Hospital, were cultivated in DMEM medium supplemented with 10% fetal calf serum.

Release of oxygen reactive species (ORS) from human neutrophils: Neutrophils (polymorphonuclears = PMN) were isolated from human blood on Ficol-Hypaque gradient and dextran sedimentation, and activated by opsonized streptococci in the presence of luminol. The luminol-induced chemiluminescence (CPM) was monitored during the reaction in a Lumac/3M Biocounter connected to a linear recorder (10).

Cell death was induced by combination of glucose oxidase (GO, producing hydrogen peroxide) trypsin and streptolysin S (SLS), and measured by the release of ^{51}Cr from prelabelled cells (4).

Phospholipase A$_2$ activity was determined by the release of radioactive arachidonic acid (ArAc) from prelabelled cells (4); cells were incubated overnight with ^{3}H-ArAc, then washed and treated, in a serum free medium, with the above inflammatory agents, and the radioactivity in the extracellular fluid was determined. Samples of this medium were analyzed by thin layer chromatography to validate that ArAc accounts for all the radioactivity in the extracellular medium after stimulation.

Lung inflammation in hamsters was induced by intratracheal (IT) administration of bleomycin (1 mg/animal) (11). For treatment with the PLA$_2$ inhibitor, the hamster received a daily intraperitoneal (IP) injection of CME (100 mg/kg) dissolved in saline, for 14 days starting one day prior to the IT administration of bleomycin. The control animals were injected with saline alone through the same procedure. The severity of the lung injury was assessed morphologically by the increase in cellularity and thickness of the intraalveolar septa, epithelial hyperplasia, and influx of inflammatory cells into the alveoli. These parameters were graded semiquantitatively to provide a morphological score on a scale of 0 (no damage) to 3 (severe damage) (12). Statistical significance was tested by the Man-Whitney test. In addition the percentage of neutrophils in the bronchoalveolar lavage fluid were determined by differential cell count.

RESULTS AND DISCUSSION

Fig. 1 depicts the effect of the PLA$_2$ inhibitor CME on the release of ORS from PMN activated by opsonized streptococci. As shown in this figure, CME inhibited the release of ORS in a dose dependent manner, and practically blocked the release at the concentration of 10 μM (1 mg/ml). The carrier alone (CMC) was ineffective.

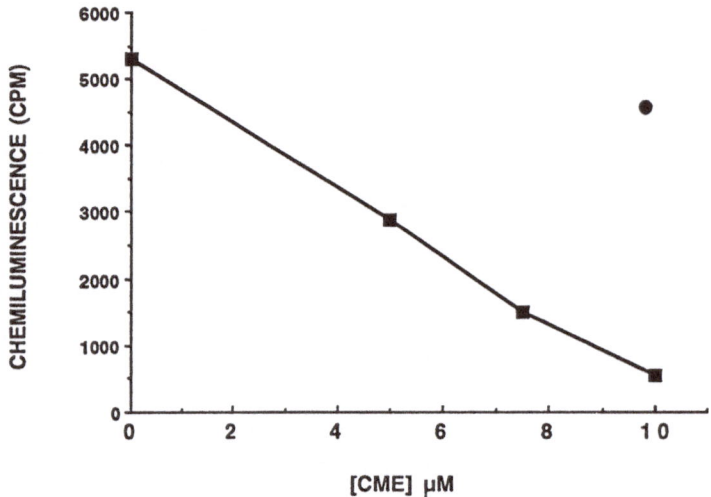

Fig. 1: **Effect of CME on the production of luminol-dependent chemiluminescence by human PMN:** Human PMN (10^6 cells/ml), were activated by opsonized strepto-cocci (STP), in the presence of luminol (5 x 10^5), and the chemiluminescence produced was monitored, as described in Experimental. For treatment with CME or CMC, these substance were added to the reaction mixture, at the indicated concentration, prior to activation with STP.   (■), CME. (●), CMC.

The effect of CME on the induction of cell death was examined, concomitantly with its effect on the release of ^{3}H-ArAc, in fibroblasts treated with combinations of inflammatory agents. This system was employed here following the report of Ginsburg et al. (4) that the process of cell killing, and concomitant release of arachidonic acid, induced by inflammatory agents (hydrogen peroxides, free radicals, hydrolytic enzymes, bacterial toxins) required their synergistic action. In the present study we have examined the effect of CME on the death of fibroblasts and release of ArAc induced by H_2O_2 (produced by GO), SLS, trypsin, and their combinations. In accord with the previous report of Ginsburg et al, a synergistic action of these substances was required to induce the cell damage. This damage was associated with activation of PLA_2 as shown by the marked enhancement of ArAc release. It was also found that the PLA_2 activity was secreted to the extracellular.

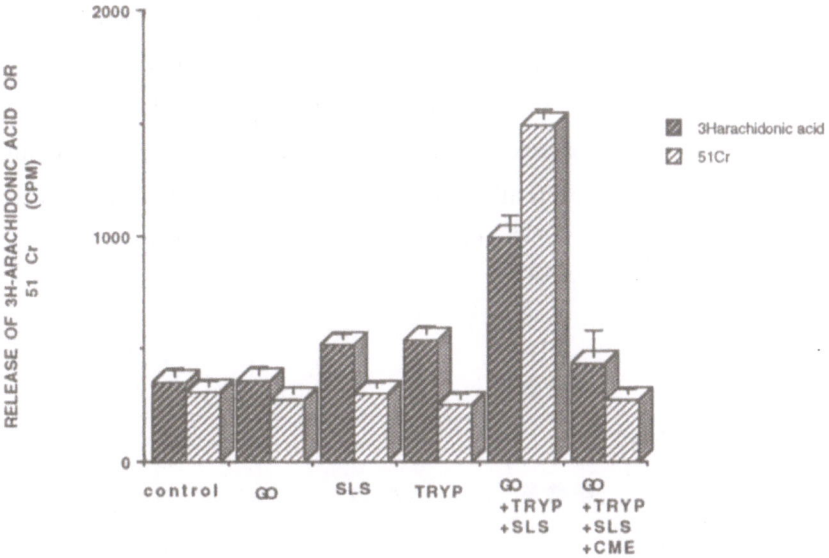

Fig 2: Effect of CME on the release of ^{51}Cr and ^{3}H-ArAc from fibroblasts stimulated with GO, SLS, trypsin and their combination: Cultured fibroblasts at confluency, prelabelled with either ^{3}H-arachidonic acid or ^{51}Cr, were washed and incubated for 1 hour with GO (2 U/ml), SLS (50 U/ml) and trypsin (250 U/ml), or their combinations as indicated. CME-treated cells were incubated with the inhibitor for 15 min prior to stimulation. The radioactivity in the extracellular medium was counted following the different treatments.

medium (not shown), and this may further contribute to the cell damage. Fig. 2 clearly shows that both the release of ^{51}Cr and of ArAc were blocked by CME.

The in vivo effect of CME on inflammatory processes was examined here in hamster lung inflammation induced by bleomycin, as this substance is known to induce membrane lipid peroxidation (13). As shown in Fig. 3, treating the hamsters with CME exerted a considerable (about 50%) reduction of the bleomycin-induced lung inflammation, as evaluated by the morphological index. More prominent was the effect of this treatment on the level of percentage of neutrophils in the bronchoalveolar lavage fluid: While in the control hamsters (treated with bleomycin + saline) the neutrophil accounted for 17.7 ± 3.2% (mean ± S.E.) of the cell counts, in the bleomycin + CME treated hamsters they accounted for only 7.3 ± 1.7%. Considering that in normal hamsters (treated with saline + saline) the basal neutrophil count was about 4%, the CME reduced the neutrophil level by about 75%. It should be noted that the treatment with CME did not induce statistically significant changes in the level of other inflammatory cells (lymphocytes and eosinophils), and this differential effect is yet to be explored. The CME-treated hamsters did not lose weight and did not show other signs of toxicity.

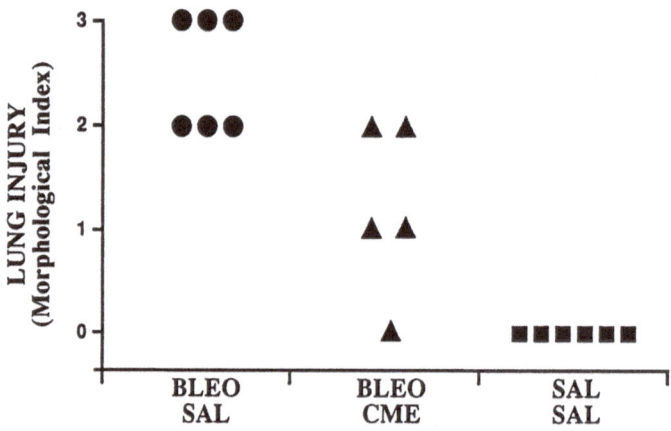

Fig. 3: **Effect of CME on bleomycin-induced lung inflammation:** Three groups of hamsters were treated as follows: IT bleomycin and IP saline (BLEO/SAL). IT bleomycin and IP CME (BLEO/CME). IT and IP saline (SAL/SAL). (See Experimental for details). The injury to the lung tissue is expressed by the morphological index described in the Experimental. BLEO/CME was statistically different from BLEO/SAL and SAL/SAL at P < 0.05 (Man-Whitney test).

The results presented above show that the cell-impermeable PLA$_2$ inhibitor, CME, is effective in controlling the different steps of the inflammatory process; it blocks the release of ORS from activated neutrophils, and the cell damage induced by inflammatory agents. As previously reported (9), similar results were obtained with monkey kidney epithelial (BGM) cells, treated with combinations of GO, SLS and a free radical generator (AAPH).

The amelioration of bleomycin-induced lung injury observed in this study is in accord with the previously reported inhibition of delayed type hypersensitivity induced in rats (9). In both studies no sign of toxicity was observed.

In another study (unpublished), a number of PLA$_2$ inhibitors of this kind, including CME, have been found to inhibits the activity of Type I (Naja Naja venom), and Type II (human recombinant synovial fluid) PLA$_2$, when applied to E. coli or lipid membranes.

It thus seems that this type of cell-impermeable PLA$_2$ inhibitors are effective in controlling the diverse actions of PLA$_2$ and related pathological conditions.

REFERENCES

1. Ginsburg I., Misgav R., Pinson R., Varani J., Ward P.A. and Kohen R. 1992. Synergism among oxidants,proteinases, hemolysins, cationic proteins and cytokines, Inflammation 16:519-538.
2. Wong P.Y.-K. and Dennis E.A. (eds). 1990. Phospholipase A$_2$: Role and function in inflammation, Plenum Press N.Y.
3. Pruzanski W. and Vadas P. 1991. Phospholipase A$_2$ - a mediator between proximal and distal effectors of inflammation. Immunology Today 12:143-146.
4. Ginsburg I., Mitra R.S., Gibbs D., Varani J. and Kohen R. 1993. The killing of endothelial cells and release of arachidonic acid: synergistic effect among hydrogen peroxides, membrane damaging agents, cationic substances and proteinases. Inflammation 17:295-319.
5. Vadas P., Browning J., Edelson J. and Pruzanski W. 1993. Extracellular phospholipase A$_2$ expression and inflammation: the relationship with associated disease states. J. Lipid Mediators 8:1-30.
6. Blackwell G.J. and Flower R.J. 1983. Phospholipase inhibition. Brit. Med. Bulletin 39:260-264.
7. Yedgar S., Reisfeld N. and Dagan A. 1986. Synthesis of cell-impermeable inhibitor of phospholipase A$_2$. FEBS Letters 200:165-168.
8. Yedgar S. and Dagan A. 1991. Phospholipase inhibiting composition and their use USA Patent No.5,064,817.
9. Yedgar S., Dagan A., Dan P. and Ginsburg I. 1994. Regulation of cell membrane phospholipase A$_2$ activity by cell-impermeable inhibitors. In: Lipid mediators in health and disease (Zor U. ed.) Freund Publishing House, Tel Aviv, pp 39-44.

10. Ginsburg I., Misgav R., Gibbs D.F., Varani J. and Kohen R. 1993. Chemiluminescence in activated human neutrophils: role of buffers and scavengers. Inflammation 17:227-243.

11. Snider G.L., Celli B.R., Goldstein R.H., O'Brien J.J. and Lucey E.C. 1978. Chronic interstitial pulmonary fibrosis produced in hamsters by endotracheal bleomycin. Am. Rev. Respir. Dis. 117:289-297.

12. Breuer R., Tochner Z., Conner M.W., Nimrod A., Gorecki M., Or R. and Slavin S. 1992. Superoxide dismutase inhibits radiation-induced lung injury in hamsters. Lung 170: 19-29.

13. Hay J., Shahzeidi S. and Laurent G. 1991. Mechanism of bleomycin-induced lung injury. Arch. Toxicol. 65:81-94.

AAS 46
Novel Molecular Approaches
to Anti-Inflammatory Theory
© 1995 Birkhäuser Verlag Basel

LEUKOCYTE ADHESION AND THE ANTI-INFLAMMATORY EFFECTS OF LEUKOCYTE INTEGRIN BLOCKADE

Thomas B. Issekutz
Department of Pediatrics
Dalhousie University
Halifax, Canada

The past several years have produced a dramatic increase in our understanding of the steps involved in the infiltration of leukocytes into inflamed tissues. At least four major families of adhesion molecules: the selectins, sialomucins, integrins, and Ig supergene family CAMs have been identified; and their interactions are being elucidated. The role of the leukocyte β_2 and α_4 integrins and the selective use of these integrins by leukocytes for migration into inflamed tissues in various organs is presented.

INTRODUCTION

Inflammation is a fundamental process in many human diseases. One of the key features of the inflammatory reaction is the infiltration of the involved tissue by blood leukocytes which dramatically alter the microenvironment of the inflamed tissue by producing cytokines, proteolytic enzymes, oxygen radicals, and frequently initiate activation of the immune system to produce sensitized T cells and specific antibody. The steps involved in the migration of leukocytes across the vascular endothelium have in recent years begun to be elucidated and found to be more complex than previously thought. It has been known for a long time that the migration of leukocytes out of the blood stream predominantly occurs at the post-capillary venules, but the endothelial cell was previously thought to play a relatively passive role. Similarly, although leukocyte activation was known to be very important for this process, the multiple stages of activation and the numerous receptors involved were not appreciated until recently.

Selectins and Sialomucins

Four major families of adhesion molecules are known to be important in this process. The selectins, which consist of three family members, L-selectin, E-selectin, and P-selectin(1,2). They are calcium dependent lectins that bind complex carbohydrates, and share a common molecular organization consisting of an N-terminal lectin domain, a domain with homology to epidermal growth factor, and 2-9 consensus repeat domains with homology to complement binding proteins, followed by a transmembrane segment and a cytoplasmic tail. L-selectin expression is limited to leukocytes being present on most neutrophils, monocytes and the majority of lymphocytes. E-selectin is expressed only on endothelial cells and appears to be limited to endothelium that has undergone activation. P-selectin is found in the α-granules of platelets and in the Weibel-Palade bodies of endothelial cells and is rapidly mobilized to the surface of these cells upon activation.

Each of these selectins are thought to bind only carbohydrates found on a family of carbohydrate rich glycoproteins known collectively as sialomucins. Sialomucins are characterized by large glycoproteins with numerous serine and threonine residues that are heavily glycosylated with O-linked carbohydrates. L-selectin has been shown to bind to the sialomucins, glycam-1 and CD34, both of which can be expressed by endothelial cells(3,4). E-selectin binds to glycoproteins containing the sialyl-Le^x and the sialyl-Le^a blood group antigens(1). There are numerous glycoproteins on leukocytes expressing these carbohydrates, including L-selectin and at least one member of the sialomucin family found on leukocytes(5). Similarly, P-selectin appears to bind to virtually identical carbohydrates as E-selectin and its major ligand on leukocytes appears to be PSGL-1, a sialomucin(6).

Integrins and Ig Family Cell Adhesion Molecules

The selectin sialomucin receptor ligand interactions are thought to mediate one phase of leukocyte extravasation. A second phase is mediated by the integrin receptors of the leukocyte binding to Ig supergene family cell adhesion molecules (CAMs) (Figure 1).

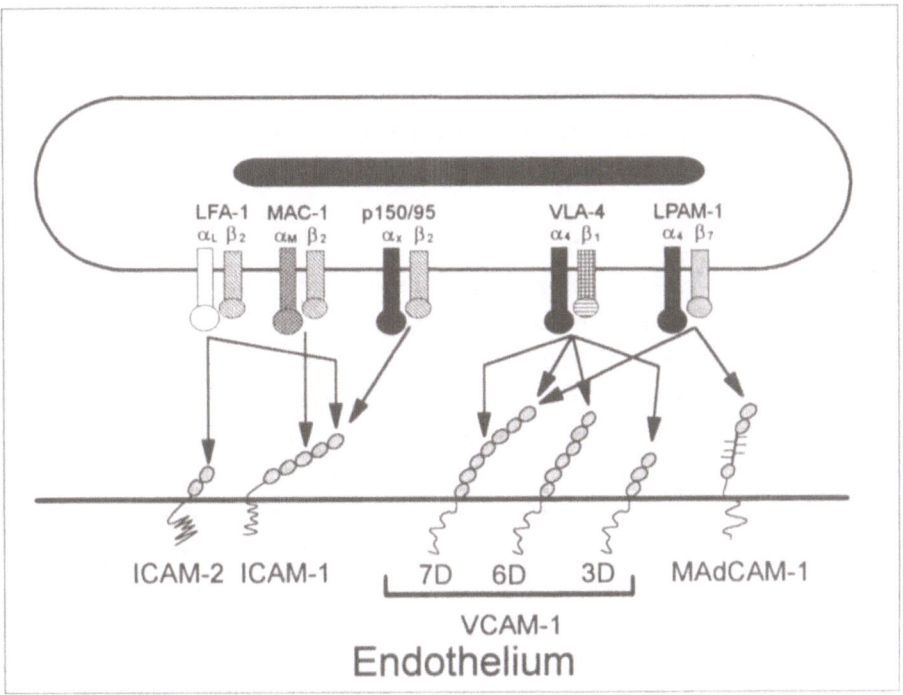

Figure 1. Diagram of a "generic" blood leukocyte showing the five leukocyte integrins and their ligands on the endothelium. The various types of leukocytes and lymphocytes express only some of these integrins, and endothelial cells only express some of the Ig family CAMs based on their tissue of origin and state of activation. (See text.)

The integrins consist of an α chain and a β chain that form a non-covalently linked heterodimer in the cell membrane(7). There are at least 14 α chains and eight β chains which combine to form at least 20 different integrin receptors. The α chains are 120-200 kDa, and most β chains are 95-130 kDa. Each of these integrins have three or more divalent cation binding sites. Five of these integrins are known to be involved in the interaction of leukocytes with vascular endothelial cells. This includes the β_2 integrins $\alpha_L\beta_2$ (CD11a/CD18) also known as lymphocyte function associated antigen-1 (LFA-1) expressed on all leukocytes, and $\alpha_M\beta_2$ (CD11b/CD18) or Mac-1, and $\alpha_X\beta_2$ (CD11c/CD18) or p150/95 found on neutrophils and monocytes. In addition, there are the two α_4 integrins $\alpha_4\beta_1$ (CD49d/CD29) or VLA-4, and $\alpha_4\beta_7$ or LPAM-1, which are found on lymphocytes and some other leukocytes.

Endothelial cells express ligands to which these integrins bind(2,7). LFA-1 binds to the Ig superfamily CAMs, ICAM-1, and ICAM-2. Mac-1 binds to ICAM-1, iC3b coated particles, fibrinogen, and denatured protein(8). P150/95 has a specificity very similar to that of Mac-1. VLA-4 binds to the Ig superfamily molecule, vascular cell adhesion molecule-1 (VCAM-1), while $\alpha_4\beta_7$ has been shown to bind to the mucosal addressin CAM or MAdCAM-1. ICAM-1 (CD54) consist of five Ig superfamily like domains composed of 90-100 amino acid loops joined by disulfide linkages. LFA-1 binds to domain 1 while Mac-1 binds to domain 3. ICAM-1 is normally expressed on endothelial cells at a low level, but is dramatically upregulated in response to cytokine activation of the endothelium. ICAM-2 (CD102) has only two extracellular domains with homology to the Ig superfamily and LFA-1 binds to domain 1 which has substantial homology with domain 1 of ICAM-1. ICAM-1 is normally expressed on endothelial cells and its expression is not affected by activation of the endothelium. VCAM-1 (CD106) has one major form that is composed of seven Ig superfamily domains. VLA-4 binds to both domains 1 and 4 of VCAM-1(9). Alternative splicing can give rise to a six domain form lacking domain 4 but this is not normally expressed, and rodents also express a three domain form. MAdCAM-1 appears to be a particularly novel hybrid molecule with three Ig superfamily domains and a sialomucin like region(10). $\alpha_4\beta_7$ binds to the N-terminal domain 1. The mucin like domain is located between domains 2 and 3 and appears to be able to mediate binding by L-selectin. MAdCAM-1 is expressed on endothelium associated with the gut associated lymphoid tissues.

The interaction of the leukocyte adhesion molecule L-selectin and integrins with their ligands also appears to involve additional factors that activate the leukocyte receptors. The affinity of

LFA-1 and Mac-1 for their ligands are greatly increased by exposure to chemotactic factors, such as IL-8.   Similarly, lymphocyte activation through the T cell receptor dramatically enhances the affinity of LFA-1 for its ligand, ICAM-1, and lymphocyte activation with phorbol esters was shown to enhance lymphocyte adhesion through VLA-4.

Multi-Step Model of Leukocyte Transendothelial Migration

This has lead to the hypothesis of a multiple step model for leukocyte endothelial cell interaction leading to extravasation (11).   The initial interaction is thought to be mediated by selectins binding to sialomucins, namely leukocyte L-selectin binding to its ligands on activated endothelium and P- and E-selectins on the endothelium binding to carbohydrates (sLe^x and sLe^a) on the leukocytes.   This interaction is unable to cause a firm adhesion of the leukocyte, but allows the leukocyte to come into close contact with the endothelium and roll on the surface of the endothelial cells.   During this rolling interaction, leukocyte receptors for chemotactic factors, including platelet activating factor, and α and β chemokines which are thought to be bound to endothelial cell surface proteoglycans activate the leukocyte. This activation is mediated through G proteins and enhances the affinity of leukocyte surface integrins which then form a strong attachment to the endothelium.   It is also associated with a spreading of the leukocyte on the surface of the endothelium and a haptotactic migration thought to be stimulated by the surface distribution of chemotactic factors(12).   The leukocyte then migrates between the endothelial cells to infiltrate the tissue. In spite of this very useful model, it is apparent from a number of investigations that this complex highly integrated sequence of events is still not fully understood.

Role of Integrins in In Vivo Neutrophil Migration

The identification of the integrins and their counter receptors have come from elegant studies of the adhesion events between leukocytes and endothelial cells in vitro.   In recent years, there has also been increasing investigation of the in vivo importance of the various integrins in leukocyte infiltration. Patients with the Leukocyte Adhesion Deficiency (LAD) syndrome lack expression of the

β_2 integrins and are unable to mobilize neutrophils to most sites of bacterial infection(13). In rabbits, anti-CD18 mAb treatment was found to reproduce this defect by strongly inhibiting neutrophil migration into cutaneous inflammatory sites induced by endotoxin, complement derived chemotactic factors (C5a), and several other agents(14). Similarly, investigations in ischemic reperfusion injury have confirmed that the blockade of the β_2 integrins decreased neutrophil infiltration and neutrophil induced tissue injury(15). However, patients with LAD are able to mobilize neutrophils to pulmonary infections, and in rabbits anti-CD18 is also unable to completely inhibit neutrophil recruitment to the inflamed lung(16). This suggests that additional CD18 independent mechanisms for neutrophil recruitment may also exist.

Our studies have further delineated the relative contributions of the two CD18 family members, LFA-1 and Mac-1, to neutrophil recruitment in the rat(17). Blockade of Mac-1 with a mAb that completely inhibits Mac-1 mediated adhesion and iC3b opsonization had no effect on neutrophil recruitment to intradermally injected $C5a_{desArg}$, IL-1α, or endotoxin. Blockade of LFA-1, on the other hand, with TA-3, a mAb to inhibit LFA-1 function, partially prevented neutrophil migration to these cutaneous sites, and the combination of blocking both Mac-1 and LFA-1, with the same antibodies virtually abolished PMN migration to $C5a_{desArg}$, IL-1, and LPS. These results suggest that LFA-1 can substitute for Mac-1 to a sufficient extent to allow neutrophil migration to proceed, but that blocking both receptors with antibodies to each of the α chains or the β chain abolishes most of the neutrophil infiltration to the inflamed skin.

The mechanism of neutrophil infiltration into other tissues appears to be quite different. In a model of rat adjuvant arthritis, neutrophil accumulation can also be readily quantified using radiolabelled blood neutrophils(18). Leukocyte accumulation increases as the degree of the arthritis progresses. Neutrophil migration to these arthritic joints is not inhibited by treatment with mAb to Mac-1, only partially inhibited with anti-LFA-1, and, unlike in skin, no further inhibition is observed by blocking both LFA-1 and Mac-1(19). Therefore neutrophil recruitment to the inflamed joint in adjuvant arthritis appears to involve a CD18 independent mechanism as well as LFA-1.

Taken together, these results suggest that the β_2 integrins form a vital adhesion pathway for neutrophil migration into inflamed skin, in ischemic reperfusion injury, and some bacterial infections, but they may play a less important role, with a CD18 independent pathway being operative, in some other types of inflammation.

Role of Integrins in In Vivo Lymphocyte Migration

The contribution of the integrins, VLA-4 and LFA-1 to lymphocyte migration in vivo, has also been examined in the last few years. Our laboratory developed a mAb, TA-2, that inhibited lymphocyte adhesion to cytokine stimulated endothelium(20). This antibody was shown to react with α_4 of VLA-4. Treatment of rats with this anti-VLA-4 mAb inhibited lymphocyte migration to cutaneous delayed type hypersensitivity (DTH) reactions, to skin injected with the cytokines, IFN-γ, IFN-α/β, and TNF-α, and to the cytokine inducers, endotoxin and poly I:C(21). The extent of inhibition, however, varied depending on the T lymphocyte population which was examined. T lymphocytes can functionally be divided into two major subpopulations, those which preferentially recirculate through peripheral lymph nodes, and are mainly virgin T cells that have not as yet encountered antigen, and "inflammatory-site-seeking" lymphocytes, which are enriched in memory T cells(22). These "inflammatory-site-seeking" T cells include lymph node lymphoblasts, obtained from antigen stimulated lymph nodes, and lymphocytes obtained from inflammatory peritoneal exudates(23). Blockade of VLA-4 inhibited up to 60% of the migration of these "inflammatory-site-seeking" lymphocytes to the cutaneous inflammatory sites, demonstrating that VLA-4 was functionally important in vivo in lymphocyte migration to inflammation(21).

Treatment of rats with mAb to LFA-1 (TA-3) was also found to block T lymphocyte migration to these same inflammatory sites(24). The effect of anti-LFA-1, however, was substantially greater on T cells obtained from the spleen, which is not enriched in the "inflammatory-site-seeking" lymphocytes, suggesting that T cells at different stages of differentiation, or from different tissue origins, utilize different integrins for infiltration into inflamed skin.

It is also interesting that the lymph node lymphoblasts and peritoneal exudate T cells, namely "inflammatory-site-seeking" T cells, used VLA-4 to a much greater extent for adhesion to cytokine activated endothelial cells than spleen T cells which employed LFA-1(25). These studies with rat lymphocytes are also in keeping with observations made using human T lymphocytes to human umbilical vein endothelial cells, which have shown that both VLA-4 and LFA-1 mediate lymphocyte adhesion to cytokine stimulated endothelium(26-28).

The effects of blocking both LFA-1 and VLA-4 have also been examined(25). Lymphocyte migration to DTH reactions induced by tuberculin protein is virtually abolished by blockade of both VLA-4 and LFA-1. Furthermore, histologically there is very little evidence of tissue infiltration by leukocytes in these reactions, since neutrophil accumulation does not occur, and lymphocyte migration appears to be completely inhibited. In addition, the characteristic induration, erythema, and, in severe lesions, haemorrhage observed in these reactions is eliminated. Interestingly, treatment with either anti-VLA-4 or anti-LFA-1 alone cannot prevent this tissue injury and is only partially active at inhibiting the migration, as noted above. This is also in agreement with the findings in patients with LAD, who lack LFA-1 but still demonstrate cutaneous DTH, apparently because of an intact VLA-4 adhesion pathway(13). Thus, VLA-4 together with LFA-1 appear to form a crucial step in lymphocyte migration into cutaneous DTH.

Inflammatory reactions mediated by T cells in other organs involve different patterns of integrin use. T lymphocyte migration to inflamed joints in rat adjuvant arthritis is one such example(29). The inflammation seeking peritoneal exudate lymphocytes migrate to the arthritic joints, but anti-VLA-4 treatment inhibits only 30-40% of this migration and anti-LFA-1 has no effect on this cell accumulation. Furthermore, spleen T cell migration to the arthritic joints is not affected by blockade of either VLA-4 or LFA-1. This is very different than observed in inflamed skin and suggests that novel pathways are involved in migration of lymphocytes into the inflamed joint.

Rat experimental allergic encephalomyelitis (EAE) is another well documented T cell induced autoimmune disease. T cell accumulation in this disease has been shown to be mediated by VLA-4(30,31). Blockade of VLA-4 can prevent the development of the neurological deficit in EAE. Thus, in the central nervous system, at least in this disease, VLA-4 is the dominant pathway.

Finally, lymphocytes also express the integrin, $\alpha_4\beta_7$, which is the homologue of the previously identified mouse integrin, $\alpha_4\beta_P$. The latter can mediate the adhesion of lymphocytes to Peyer's patch high endothelial venules and is thought to act through its ability to bind to MAdCAM on these cells(32,33). The TA-2 mAb reacts with α_4 both on $\alpha_4\beta_1$ and $\alpha_4\beta_7$. Blockade of α_4 in vivo has dramatically inhibited lymphocyte migration to both Peyer's patches and mesenteric lymph nodes(21). T cell migration to Peyer's patches was inhibited by >98%, while migration to mesenteric lymph nodes was inhibited by 80%. This strongly implies that α_4 integrins, mostly likely $\alpha_4\beta_7$ and its counter receptor MAdCAM, mediate an important step in the process of lymphocyte homing to these two intestinal lymphoid tissues.

Recent studies have also suggested that α_4 integrins have an important role in intestinal inflammation(34). Rats infected with Trichinella spiralis develop a T cell dependent mucosal immune response to this parasite that results in its expulsion after about two weeks. Treatment of animals with antibody to α_4 inhibited the expulsion of the trichinella. The migration of immune T cells developing in the mesenteric lymph nodes in response to the nematode were inhibited from entering the thoracic duct, and the ability of immune T cells to localize in the inflamed lamina propria was dramatically reduced (>95%). Thus, α_4 integrin blockade also strongly suppressed T cell dependent responses in the gut. Treatment of rats with anti-LFA-1 did not have an effect on T. Spiralis expulsion and localization of the intestine.

In summary, these studies demonstrate that the leukocyte integrins form a key step in neutrophil and lymphocyte migration out of the blood and infiltration into tissues. The in vivo studies also suggest that, although there is a lot of redundancy in integrin usage by leukocytes for tissue infiltration, both the mechanism of inflammation and the organ involved play a major role in determining the integrins utilized by leukocytes for adhesion to the vascular endothelium and their transmigration.

REFERENCES

1. Lasky LA. Selectins: Interpreters of cell-specific carbohydrate information during inflammation. Science 1992; 258:964-969.

2. Springer TA. Traffic signals for lymphocyte recirculation and leukocyte emigration: The multistep paradigm. Cell 1994; 76:301-314.

3. Imai Y, Lasky LA, Rosen SD. Sulphation requirement for GlyCAM-1, an endothelial ligand for L-selectin. Nature 1993; 361:555-557.

4. Baumhueter S, Singer MS, Henzel W, et al. Binding of L-selectin to the vascular sialomucin CD34. Science 1993; 262:436-438.

5. Levinovitz A, Mühlhoff J, Isenmann I, Vestweber D. Identification of a glycoprotein ligand for E-selectin on mouse myeloid cells. J Cell Biol 1993; 121:449-459.

6. Moore KL, Stults NL, Diaz S, et al. Identification of a specific glycoprotein ligand for P-selectin (CD62) on myeloid cells. Journal of Cell Biology 1992; 118:445-456.

7. Sonnenberg A. Integrins and thier ligands. Current Topics in Microbiology and Immunology 1993; 184:7-35.

8. Diamond MS, Garcia-Aguilar J, Bickford JK, Corbi AL, Springer TA. The I domain is a major recognition site on the leukocyte integrin Mac-1 (CD11b/CD18) for four distinct adhesion ligands. J Cell Biol 1993; 120:1031-1043.

9. Vonderheide RH, Tedder TF, Springer TA, Staunton DE. Residues within a conserved amino acid motif of domains 1 and 4 of VCAM-1 are required for binding to VLA-4. J Cell Biol 1994; 125:215-222.

10. Briskin MJ, McEvoy M, Butcher EC. MAdCAM-1 has homology to immunoglobulin and mucin-like adhesion receptors and to IgA1. Nature 1993; 363:461-464.

11. Butcher EC. Leukocyte-endothelial cell recognition: Three (or more) steps to specificity and diversity. Cell 1991; 67:1033-1036.

12. Rot A. Endothelial cell binding of NAP-1/IL-8:   Role in neutrophil emigration. Immunol Today 1992; 13:291-294.

13. Anderson DC, Schmalsteig FC, Finegold MJ, et al. The severe and moderate phenotypes of heritable Mac-1, LFA-1 deficiency: Their quantitative definition and relation to leukocyte dysfunction and clinical features. J Infect Dis 1985; 152:668-689.

14. Nourshargh S, Rampart M, Hellewell PG, et al. Accumulation of 111In-neutrophils in rabbit skin in allergic and non-allergic inflammatory reactions in vivo.   Inhibition by neutrophil pretreatment in vitro with a monoclonal antibody recognizing the CD18 antigen. J Immunol 1989; 142:3193-3198.

15. Vedder NB, Winn RK, Rice CL, Chi EY, Arfors K-E, Harlan JM. Inhibition of leukocyte adherence by anti-CD18 monoclonal antibody attenuates reperfusion injury in the rabbit ear. Proc Natl Acad Sci USA 1990; 87:2643-2646.

16. Doerschuk CM, Winn RK, Coxson HO, Harlan JM. CD18-dependent and -independent mechanisms of neutrophil emigration in the pulmonary and systemic microcirculation of rabbits. J Immunol 1990; 144:2327-2333.

17. Issekutz AC, Issekutz TB. The contribution of LFA-1 (CD11a/CD18) and MAC-1 (CD11b/CD18) to the in vivo migration of polymorphonuclear leukocytes to inflammatory reactions in the rat. Immunol 1992; 76:655-661.

18. Issekutz AC, Issekutz TB. Quantitation and kinetics of polymorphonuclear leukocyte and lymphocyte accumulation in joints during adjuvant arthritis in the rat. Lab Invest 1991; 64:656-663.

19. Issekutz AC, Issekutz TB. A major portion of polymorphonuclear leukocyte and T lymphocyte migration to arthritic joints in the rat is via LFA-1/MAC- 1-independent mechanisms. Clin Immunol Immunopathol 1993; 67:257-263.

20. Issekutz TB, Wykretowicz A. Effect of a new monoclonal antibody, TA-2, that inhibits lymphocyte adherence to cytokine stimulated endothelium in the rat. J Immunol 1991; 147:109-116.

21. Issekutz TB. Inhibition of in vivo lymphocyte migration to inflammation and homing to lymphoid tissues by the TA-2 monoclonal antibody: A likely role for VLA-4 in vivo. J Immunol 1991; 147:4178-4184.

22. Cush JJ, Pietschmann P, Oppenheimer-Marks N, Lipsky PE. The intrinsic migratory capacity of memory T cells contributes to their accumulation in rheumatoid synovium. Arthritis Rheum 1992; 35:1434-1444.

23.   Issekutz TB, Webster DM, Stoltz JM. Lymphocyte recruitment in
      vaccinia   virus-induced   cutaneous   delayed-type
      hypersensitivity. Immunol 1986; 58:87-94.

24.   Issekutz TB. Inhibition of lymphocyte endothelial adhesion and
      in vivo lymphocyte migration to cutaneous inflammation by
      TA-3, a new monoclonal antibody to rat LFA-1. J Immunol 1992;
      149:3394-3402.

25.   Issekutz TB. Dual inhibition of VLA-4 and LFA-1 maximally
      inhibits cutaneous delayed-type hypersensitivity-induced
      inflammation. Am J Pathol 1993; 143:1286-1293.

26.   Haskard D, Cavender D, Beatty P, Springer T, Ziff M. T
      lymphocyte adhesion to endothelial cells: mechanisms
      demonstrated by anti-LFA-1 monoclonal antibodies. J Immunol
      1986; 137:2901-2906.

27.   Oppenheimer-Marks N, Davis LS, Bogue DT, Ramberg J, Lipsky PE.
      Differential utilization of ICAM-1 and VCAM-1 during the
      adhesion   and   transendothelial   migration   of   human   T
      lymphocytes. J Immunol 1991; 147:2913-2921.

28.   Rice GE, Munro JM, Bevilacqua MP. Inducible cell adhesion
      molecule 110 (INCAM-110) is an endothelial receptor for
      lymphocytes. A CD11/CD18-independent adhesion mechanism. J
      Exp Med 1990; 171:1369-1374.

29.   Issekutz TB, Issekutz AC. T lymphocyte migration to arthritic
      joints and dermal inflammation in the rat: Differing
      migration patterns and the involvement of VLA-4. Clin Immunol
      Immunopathol 1991; 61:436-447.

30.   Yednock TA, Cannon C, Fritz LC, Sanchez-Madrid F, Steinman L,
      Karin   N.   Prevention   of   experimental   autoimmune
      encephalomyelitis by antibodies against $\alpha 4\beta 1$ integrin. Nature
      1992; 356:63-66.

31.   Issekutz TB. Effect of anti-LFA-1 and anti-VLA-4 on T
      lymphocyte migration to skin, joint, and CNS inflammation and
      lymph nodes. Int Cong Immunol 1992; 8:288.(abstract)

32.   Holzmann B, McIntyre BW, Weissman IL. Identification of a
      murine Peyer's patch-specific lymphocyte homing receptor as an
      integrin molecule with an α chain homologous to human VLA-4α.
      Cell 1989; 56:37-46.

33.   Berlin C, Lang EL, Briskin MJ, et al. $\alpha 4\beta 7$ integrin mediates
      lymphocyte binding to the mucosal vascular addressin MAdCAM-1.
      Cell 1993; 74:1-20.

34.   Bell RG, Issekutz T. Expression of a protective intestinal
      immune response can be inhibited at three distinct sites by
      treatment with anti- $\alpha 4$ integrin. J Immunol 1993;
      151:4790-4802.

AAS 46
Novel Molecular Approaches
to Anti-Inflammatory Theory
© 1995 Birkhäuser Verlag Basel

ANTI-ICAM in Inflammatory Disease

Robert Rothlein, Ph. D.

Boehringer Ingelheim, 900 Ridgebury Road,
Ridgefield, CT 06877, USA

INTRODUCTION

The anti-inflammatory activity of a monoclonal antibody to intercellular adhesion molecule-1 (ICAM-1) has been evaluated in vitro, in vivo and in the clinic.

ICAM-1 is a broadly distributed adhesion molecule and a member of the immunoglobulin supergene family (1-10). It is a ligand for at least two members of the CD18 family of adhesion molecules as well as CD43 and fibrinogen (11,12). When ICAM-1 was identified as the ligand of LFA-1, it was thought to be functional only in lymphocyte mediated processes (1,13). Subsequently it was found to also mediate neutrophil adhesion to and migration through endothelium (14,15). Finally, antibodies to ICAM-1 have been shown to inhibit both lymphocyte and granulocyte adhesion processes as shown

by functional in vitro assays such as MLRs (16),
antigen-induced proliferation (17), cytotoxic T-cell
activity (18,19), and granulocyte and lymphocyte
attachment to endothelium (14,15,20).

Since antibodies to ICAM-1 have inhibited in
vitro models of inflammation, the in vivo activity of
antibodies to ICAM-1 in models of neutrophil and/or
lymphocyte mediated inflammatory disease was
evaluated. One of the earliest experiments to be
performed were those by Barton et. al. demonstrating
that anti-ICAM-1 MAb inhibited neutrophil influx into
rabbit lungs following systemic activation with
phorbol esters (21), providing in vivo evidence that
anti-ICAM-1 MAb did indeed block neutrophil function.
Antibodies to ICAM-1 also mitigated eosinophil influx
and airway hyperresponsiveness in a non human primate
model of antigen- induced airway hyperresponsiveness
(9), kidney allograft rejection and reversal of acute
rejection in a non human primate model of solid organ
transplantation (8), neurological damage in a rabbit
model of spinal cord ischemia/reperfusion and stroke
(22-24), joint inflammation in a rat model of adjuvant
induced arthritis (25) and loss of blood flow to
marginal zones surrounding full thickness burns in a
rabbit model of burn (26).

Anti-ICAM-1 in Man

The findings thus far described support a role for the use of anti-ICAM-1 in a number of diseases. The specific anti-ICAM-1 monoclonal antibody selected for clinical trials was an IgG2a which binds to domain 2 of ICAM-1. This antibody, called BIRR1 (also called R6.5), blocks both the CD11a,CD18 and the CD11b,CD18 component to lymphocyte and neutrophil adhesion (27,28).

BIRR1 was first tested in man, as a proof of concept, in solid organ transplantation and rheumatoid arthritis. Phase I-IIa studies in stroke and burn are ongoing. Although BIRR1 is a murine monoclonal antibody and, as such, is immunogenic in man, early evidence suggests that it is effective and relatively safe.

Solid organ transplant was the first clinical indication where BIRR1 was tested. Phase I-IIa has been completed for renal transplant and a phase IIb study that is double blind and placebo controlled is ongoing. A Phase I-IIa trial in liver transplantation has also been completed.

The phase I-IIa renal transplantation studies involved cadaveric renal allograft recipients at "high" risk for post transplant complications such as delayed graft function (DGF, a complication of renal transplant with an ischemia/reperfusion component) and acute rejection. These patients were given BIRR1 prophylactically as an add-on medication to conventional triple therapy immunosuppression (azathioprine, steroids, and cyclosporin A). The primary objectives of this study were to evaluate the safety and pharmacokinetics of BIRR1 as well as to get a signal as to BIRR1's ability to reduce the incidence of DGF and acute rejection (29).

Ten patients were initially intended to be enrolled in the trial, 5 each on two different dosing regimens. These dosing regimens were direct extrapolations from those which were efficacious in the cynomolgus monkeys, as well as the investigator's experience with OKT_3. It was anticipated that both of these regimens would provide BIRR1 levels in a potentially therapeutic range.

Early results from this open labeled study revealed that the original dosing projections were too low to attain target serum concentrations of BIRR1. This target level was established by determining the minimum concentration of BIRR1 necessary to achieve maximal inhibition of adhesion in vitro. Thus, in order to fulfill the objective of the study (i.e. to find two dosing regimens which achieved the pharmacokinetic target), a total of 5 dosing regimens were studied in 18 patients ranging from a total of 150mg of antibody over 14 day period to a total of

560mg of antibody over a 6 day period. Of the five dosing regimens studied, the two highest regimens consistently met the pharmacokinetic target. Both were 6 day regimens: 160 mg pre transplant loading, followed by 40 mg/day x 5 days (abbreviated as 160/40 x 5), and a 160/80 x 5 regimen, for a total dose of 360mg and 560 mg, respectively.

Although there is no hard data to support this, it is felt that three potential reasons for the increased dosing requirement for BIRR1 in renal allograft recipients when compared to animal studies may have to do with more circulating ICAM-1 in humans when compared to other species; a higher affinity and avidity of BIRR1 for human ICAM-1 than other species ICAM-1 and/or sick human beings have more intrinsic ICAM-1 expression due to the underlying disease than do healthy non-human primates.

The study design only allowed for efficacy comparisons with a historical control. However comparisons between groups achieving target blood concentrations vs. those that did not achieve target blood levels became possible as the study evolved. Altogether, 18 patients were enrolled receiving 5 different dosing regimens. Eleven patients received anti-ICAM-1 regimens which achieved the pharmacokinetic target; seven did not. The outcome of these two groups were compared.

The incidence of both acute rejection (followed out to three months) and DGF (a shorter-term endpoint evaluated by the end of the first post transplant week) was lower in the 11 patients with adequate BIRR1 levels. In fact, of the 7 patients not achieving pharmacokinetic target levels of BIRR1, all had at least one rejection episode within the first three months post-transplant while only 3 of 11 patients achieving target levels of BIRR1 had a rejection episode within the first three months post-transplantation. In terms of DGF, anti-ICAM-1 also seemed to be protective since 6/7 of those patients receiving sub therapeutic levels of BIRR1 had DGF while 6/11 of those patients receiving therapeutic levels had DGF.

Results from liver transplant trials also were encouraging. Like the kidney transplant Phase I-IIa trial, anti-ICAM-1 was given for 6 days as an add-on therapy to traditional triple therapy at the time of transplantation. However, liver transplant patients required more anti-ICAM-1 to achieve pharmacokinetic target blood levels than did the renal transplant patients. In fact the highest 6 day dosing regimen in liver transplant entailed a total 6 day dose of 880mg compared to 560mg in renal transplant patients. Overall, the safety profile of BIRR1, even at the highest dosing regimen, was good (30).

Rheumatoid Arthritis (RA)

Anti-ICAM-1 was also evaluated in an investigator's IND study in an open-label, dose escalating phase I-IIa study in patients with severe RA by Dr. Peter Lipsky (University of Texas, Southwestern Medical Center, Dallas) (31). Although primarily a PK and safety study, preliminary efficacy was also assessed using the Paulus criteria described below. The population under study included patients with active disease of greater than 4 years (average 18 years) duration who were functional class I-III and who had failed at least 2 disease-modifying drug (average 4 drugs. The patient were allowed to be on nonsteroidal anti-inflammatory drugs (NSAIDS) or $\leq$10mg prednisone/day during active treatment with BIRR1. Twenty-three patients completed a full five day course of treatment with dosing regimens ranging from 60/20mg x 4 days to 240/80mg x 4 days with the majority of the patients receiving a regime of 120/40mg x 4 days. In all of the dosing regimens studied, there have been no serious adverse events however some of the patients (particularly those receiving the higher loading doses) had a mild to moderate "first dose reaction". The pharmacokinetic target was achieved in all patients receiving the 120/40mg x4 day regimen.

Efficacy was assessed in this study using the modified Paulus criteria, which examines the following

six parameters including joint swelling, joint
tenderness, morning stiffness, patient/physician
global assessment, and erythrocyte sedimentation rate
(ESR). In order to be classified as a "responder" a
patient must have had a $\geq$ 20% reduction in 4/6
parameters.

Results to date reveal that of the 23 patients
receiving full treatment, 13 were considered
responders with 9 showing a response up to 60 days
post-treatment. Responses appeared to wane between 60-
90 days post-treatment.

Thus, in severe RA, BIRR1 has been tested at
several dosing regimens. The 120/40mg x 4 days dosing
regimen (280 mg total) achieved the pharmacokinetic
target, and resulted in a clinical response in more
than half the patients for a prolonged period of time.
Furthermore, during treatment, T cell counts in the
blood appear to be elevated but rapidly return to
normal levels upon cessation of therapy.

In conclusion, the inhibition of leukocyte
adhesion by antagonizing CD18/ICAM-1 interactions
appears to be a viable therapeutic approach to
attenuate the inflammatory response. Antibodies to
ICAM-1 offer a good first generation product to
achieve CD18/ICAM-1 antagonism.

REFERENCE
1. Dustin, M. L., R. Rothlein, A. K. Bhan, C. A. Dinarello, and T. A. Springer. 1986. Induction by IL-1 and interferon, tissue distribution, biochemistry, and function of a natural adherence molecule (ICAM-1). J. Immunol. 137:245.

2. Dustin, M. L., K. H. Singer, D. T. Tuck, and T. A. Springer. 1988. Adhesion of T lymphoblasts to epidermal keratinocytes is regulated by interferon gamma and is mediated by intercellular adhesion molecule-1 (ICAM-1). J. Exp. Med. 167:1323.

3. Pober, J. S., M. A. Gimbrone Jr., L. A. Lapierre, D. L. Mendrick, W. Fiers, R. Rothlein, and T. A. Springer. 1986. Overlapping patterns of activation of human endothelial cells by interleukin 1, tumor necrosis factor and immune interferon. J. Immunol. 137:1893.

4. Frohman, E. M., T. C. Frohman, M. L. Dustin, B. Vayuvegula, B. Choi, A. Gupta, S. van den Noort, and

S. Gupta. 1989. Induction of ICAM-1 expression on human fetal astrocytes by interferon-gamma, tumor necrosis factor-alpha and interleukin-1: relevance to intracerebral antigen presentation. J. Neuroimmunol. 23:117.

5. Vejlsgaard. G. L., E. Ralfkiaer, C. Avnstorp, M. Czajkowski, S. D. Marlin, and R. Rothlein. 1989. Kinetics and characterization of intercellular adhesion molecule-1 (ICAM-1) expression on keratinocytes in various inflammatory skin lesions and malignant cutaneous lymphomas. J. Amer. Acad. Dermatol. 20:782.

6. Griffiths, C. E. M., and B. J. Nickoloff. 1989. Keratinocyte intercellular adhesion molecule-1 (ICAM-1) expression precedes dermal T lymphocytic infiltration in allergic contact dermatitis (Rhus dermatitis). Am. J. Pathol. 135:1045.

7. Adams, D. H., S. G. Hubscher, J. Shaw, R. Rothlein, and J. M. Neuberger. 1989. Intercellular adhesion molecule 1 on liver allografts during rejection. Lancet 2:1122.

8. Cosimi, A. B., D. Conti, F. L. Delmonico, F. I. Preffer, S. -L. Wee, R. Rothlein, R. Faanes, and R. B. Colvin. 1990. In vivo effects of monoclonal antibody to ICAM-1 (CD54) in nonhuman primates with renal allografts. J. Immunol. 144:4604.

9. Wegner, C. D., R. H. Gundel, P. Reilly, N. Haynes, L. G. Letts, and R. Rothlein. 1990. Intercellular adhesion molecule-1 (ICAM-1) in the pathogenesis of asthma. Science 247:456.

10. Matsui, M., M. Temponi, and S. Ferrone. 1987. Characterization of a monoclonal antibody-defined human melanoma-associated antigen susceptible to induction by immune interferon. J. Immunol. 139:2088.

11. Rosenstein, Y., J. K. Park, W. C. Hahn, F. S. Rosen, B. E. Bierer, and S. J. Burakoff. 1991. CD43, a molecule defective in Wiskott-Aldrich syndrome, binds ICAM-1. Nature 354:233.

12. Languino, L. R., J. Plescia, A. Duperray, A. A. Brian, E. F. Plow, J. E. Geltosky, and D. C. Altieri.

1993. Fibrinogen mediates leukocyte adhesion to vascular endothelium through an ICAM-1-dependent pathway. Cell 73:1423.

13. Rothlein, R., M. L. Dustin, S. D. Marlin, and T. A. Springer. 1986. A human intercellular adhesion molecule (ICAM-1) distinct from LFA-1. J. Immunol. 137:1270.

14. Smith, C. W., R. Rothlein, B. J. Hughes, M. M. Mariscalco, F. C. Schmalstieg, and D. C. Anderson. 1988. Recognition of an endothelial determinant for CD18 - dependent neutrophil adherence and transendothelial migration. J. Clin. Invest. 82:1746.

15. Smith, C. W., S. D. Marlin, R. Rothlein, C. Toman, and D. C. Anderson. 1989. Cooperative Interactions of LFA-1 and Mac-1 with Intercellular Adhesion Molecule-1 in Facilitating Adherence and Transendothelial Migration of Human Neutrophils in vitro. J. Clin. Invest. 83:2008.

16. Boyd, A. W., S. O. Wawryk, G. F. Burns, and J. V. Fecondo. 1988. Intercellular adhesion molecule 1 (ICAM-1) has a central role in cell-cell contact-mediated immune mechanisms. Proc. Natl. Acad. Sci. U. S. A. 85:3095.

17. Dougherty, G. J., S. Murdoch, and N. Hogg. 1988. The function of human intercellular adhesion molecule-1 (ICAM-1) in the generation of an immune response. Eur. J. Immunol. 18:35.

18. Mentzer, S. J., R. Rothlein, T. A. Springer, and D. V. Faller. 1988. Intercellular adhesion molecule-1 (ICAM-1) is involved in the cytolytic T lymphocyte interaction with human synovial cells. J. Cell. Physiol. 137:173.

19. Makgoba, M. W., M. E. Sanders, G. E. G. Luce, E. A. Gugel, M. L. Dustin, T. A. Springer, and S. Shaw. 1989. Intercellular Adhesion Molecule-1 (ICAM-1) monoclonal antibody inhibits cytotoxic T lymphocyte recognition. Ann. NY Acad. Sci. 532:427.

20. Dustin, M. L., and T. A. Springer. 1988. Lymphocyte function associated antigen-1 (LFA-1) interaction with intercellular adhesion molecule-1 (ICAM-1) is one of at least three mechanisms for lymphocyte adhesion to cultured endothelial cells. J. Cell Biol. 107:321.

21. Barton, R. W., R. Rothlein, J. Ksiazek, and C. Kennedy. 1989. The effect of anti-intercellular adhesion molecule-1 on phorbol-ester-induced rabbit lung inflammation. J. Immunol. 143:1278.

22. Clark, W. M., K. P. Madden, R. Rothlein, and J. A. Zivin. 1991. Reduction of central nervous system ischemic injury in rabbits using leukocyte adhesion antibody treatment. Stroke 22:877.

23. Clark, W. M., K. P. Madden, R. Rothlein, and J. A. Zivin. 1991. Reduction of central nervous system ischemic injury by monoclonal antibody to intercellular adhesion molecule. J. Neurosurg. 75:623.

24. Bowes, M. P., J. A. Zivin, and R. Rothlein. 1993. Monoclonal antibody to the ICAM-1 adhesion site reduces neurological damage in a rabbit cerebral embolism stroke model. Exp. Neurol. 119:215.

25. Iigo, Y., T. Takashi, T. Tamatani, M. Miyasaka, T. Higashida, H. Yagita, K. Okumura, and W. Tsukada. 1991. ICAM-1-dependent pathway is critically involved in the pathogenesis of adjuvant arthritis in rats. J. Immunol. 147:4167.

26. Mileski, W. J., D. Borgstrom, E. Lightfoot, R. Rothlein, R. Faanes, P. E. Lipsky, and C. Baxter. 1992. Inhibition of leukocyte-endothelial adherence following thermal injury. J. Surg. Res. 52:334.

27. Staunton, D. E., M. L. Dustin, H. P. Erickson, and T. A. Springer. 1990. The arrangement of the immunoglobulin-like domains of ICAM- 1 and the binding sites for LFA-1 and rhinovirus. Cell 61:243.

28. Diamond, M. S., D. E. Staunton, A. R. De Fougerolles, S. A. Stacker, J. Garcia-Aguilar, M. L. Hibbs, and T. A. Springer. 1990. ICAM-1 (CD54): A counter-receptor for Mac-1 (CD11b/CD18). J. Cell Biol. 111:3129.

29. Haug, C. E., R. B. Colvin, F. L. Delmonico, H. Auchincloss,Jr., N. Tolkoff-Rubin, F. I. Preffer, R. Rothlein, S. Norris, L. Scharschmidt, and A. B. Cosimi. 1993. A phase I trial of immunosuppression with anti-ICAM-1 (CD54) mAb in renal allograft recipients. Transplantation 55:766.

30. Davies, M. H., G. A. Tregunno, L. A. Scharschmidt, and J. M. Neuberger. 1993. Monoclonal anti-ICAM-1 antibodies in liver transplantation: Preliminary results of phase I trial. Hepatology 18:75a.(Abstract)

31. Kavanaugh, A. F., L. S. Davis, l. a. Nichols, S. H. Norris, R. Rothlein, L. A. Scharschmidt, and P. E. Lipsky. 1994. Treatment of refractory rheumatoid arthritis with a monoclonal antibody to intercellular adhesion molecule 1. Arthritis Rheum. 37:992.

AAS 46
Novel Molecular Approaches
to Anti-Inflammatory Theory
© 1995 Birkhäuser Verlag Basel

COMPLEMENTARY PEPTIDES AS RECOGNITION MOLECULES

Giorgio Fassina

Protein Engineering , TECNOGEN S.C.p.A.
Parco Scientifico, 81015 Piana di Monte Verna, (CE), ITALY

SUMMARY.   The possibility of designing sequence-directed recognition peptides (complementary peptides) able to non covalently associate target peptides or proteins is one of the most important applications deriving from the Molecular Recognition Theory [MRT]. Complementary peptides can be used widely not only as synthetic ligands for the development of affinity purification strategies to isolate target peptides or proteins from crude sources, but more importantly as peptidyl antagonists to inhibit biologically relevant interactions, or to probe functional sites in proteins and corresponding receptors.

INTRODUCTION

The current knowledge of the mechanisms involved in recognition processes is not sufficient to allow a simple and easy design of molecules able to interact with target polypeptides, and therefore usable as inhibitors of biologically relevant interactions. In the past decade, a novel approach has been proposed to predict peptidyl ligands with potential therapeutic applications. It has been observed experimentally [1] that peptides deduced from the transcription of complementary DNA strands were characterised by recognition properties for each other. A main feature of these peptide pairs is their hydropathic anti-complementarity, since complementary strands of DNA encode amino acid residues that are characterised by an inversion of hydropathy of their side chain groups [2].   Specifically, hydrophobic residues are complemented by hydrophilic residues and *vice versa*, while slightly hydrophilic residues are complemented by slightly hydrophobic residues.   This relationship was shown to exist when codons are read either from the conventional 5'-3' direction, or from the 3'-5' direction, since is the codon middle base which characterise the hydropathy of the amino acid residue.   The role of complementary hydropathy for producing binding has been validated by the observation that complementary peptides could be designed starting from the target sequence using a computer assisted method [AMINOMAT] to predict the best complementary sequence, without prior knowledge of the nucleotide sequence coding for the target [3].   Additional evidences on the prominent role of the hydropathic characteristics of the peptides rather than their amino acid sequence are provided by the observation that once the hydropathic profile of the complementary peptide is maintained, recognition properties are maintained. Complementary peptide deduced from 5'-3' or 3'-5' translation of complementary nucleotide sequences bind in most cases equally well their target sequence even if their amino acid sequence is different, since both peptides display a very similar hydropathic profile [4].   Sequence inversion of complementary peptides leads also to retention of recognition properties [5], because also in this case the peptide hydropathic profile results unaltered, once is aligned in an antiparallel way with the parent peptide.

More recently, the synthetic procedure used for the production of Multiple Antigenic Peptides [MAP] [6] has been applied to the synthesis of complementary peptides, starting from polydentate lysine core with four or eight complementary peptide chains attached [7]. Multimeric peptides can be immobilised directly on solid supports for the preparation of affinity columns [7,8], and display several advantages in comparison with the immobilisation of their linear counterparts.   First, following immobilisation only a limited number of peptide chains of the multimer result covalently linked to the solid phase , leaving the remainder facing the mobile phase and sufficiently spaced to interact properly with the target [8].   Second, ligand multimerization generally enhances the interaction avidity in the case of polyvalent targets [9].   Spacer length, incorporation of solubilizing residues and a reduced or augmented number of chains linked to the central core can be selected according to the characteristics of the peptidyl ligand under consideration.

The basic ideas sustaining the MRT can be applied also to the identification of protein sites involved in receptor recognition, and this is of extreme importance to study the protein functional role and in the design of novel molecules acting as antagonists or agonists of the protein function.   Given a protein known to interact with a receptor, several methodologies can be used in order to identify which part of the molecule is important for recognition.   When the structure of the protein/receptor complex is not available, as it is often the case, site directed mutagenesis, chemical modification of specific residues, limited proteolysis and use of antibodies or synthetic peptides, alone or in combination, can provide useful indications about the protein regions important for binding. In addition to these methodologies , a corollary of the Molecular Recognition Theory [MRT] [1] suggested the possibility to search for regions of hydropathic complementarity between protein and   corresponding receptor sequences. According to MRT, complementary peptides to peptide ligands should have similar conformation to the binding site of the receptor of the ligand and therefore similar sequence. Since complementary peptides are deduced from complementary DNA strands [2], and since a codon that specifies a hydrophilic residue has a complementary codon that specifies a hydrophobic residue, and *vice versa*, complementary nucleotide sequences encode peptides with inverted hydropathic profile.   Consequently, protein/receptor binding sites should be characterised by hydropathic anti-complementarity.   In many cases where the nucleotide sequences corresponding to   proteins and receptors were known, as well as the precise localisation of the binding sites, a striking hydropathic correlation has been found between these sites.   Examples in which this approach has been used successfully to determine ligand receptor interaction sites include human follicle stimulating hormone and its receptor, bombesin and its receptor , interleukin 2 and its receptor , ferritin and its receptor , cystatin-C4, G-protein coupled receptors and their hormones, hematopoietin receptors and their hormones, tyrosine kinase growth factor receptors and their receptors [10].   In order to further test the applicability of this approach for predicting protein/receptor binding sites, a computer program, named SITESEARCH [11] has been developed,   able to compare the hydropathic profiles of the protein and receptor sequences under investigation, and to determine the sequence stretches characterised by the highest level of hydropathic complementarity.   The program uses the Kyte and Doolittle algorithm for calculating the hydropathic profile of each segment of predefined length L (generally between 5 and 11 residues) in the two chains under examination. For each segment pair of length L, the program determines also the hydropathic complementarity score Θ, which is and indicator of the degree of complementarity.   Lower the score, highest the hydropathic complementarity. In other words, the programs scans the receptor and protein sequences simultaneously to find the highest complementary sites.

This paper will review applications of the MRT to two proteinaceous factors involved in inflammation,   lipocortin 1 and interleukin-1β, discussing in the first case the design and characterisation of a complementary peptide able to recognise a selected fragment of lipocortin 1, while in the second case will be presented the identification of the putative binding site between interleukin-1β and its type I receptor .

METHODS

Computer assisted design of hydropathically complementary peptides.

Selection of hydropathically complementary sequences to target sequences is carried out using a mathematical approach similar to that developed by Kyte & Doolittle [12] for displaying the hydropathic profile of polypeptide sequences.   Briefly, to each residue of the target sequence is assigned a set of hydropathically complementary amino acids, with opposite hydropathy values in the range of + 1.0 hydropathy units, except for threonine, where a +1.1 hydropathy range is used.

The moving average hydropathy a_i of the target sequence is then calculated according to the formula.

$$a_k = \sum_{i=k-s}^{k+s} h_i$$

where h_i is the hydropathy value of each residue A_i of the target sequence, k is (s) to (n-s), and s is (r-1)/2, where r, averaging window, is an odd numeral up to or equal n, the number of residues in the target sequence. In a similar manner, the moving average hydropathy b_i of all the complementary sequences is determined according to the expression:

$$b_k = \sum_{i=k-s}^{k+s} g_{i,j}$$

where $g_{i,j}$ correspond to the hydropathy value of the complementary residue. Given an averaging window r, the complementary sequence characterised by the lowest score Θ, which is inversely proportional to the degree of complementarity on a moving average base and defined as:

$$\Theta = \sum \sqrt{(a_i+b_i)^2/(n-2s)}$$

is selected as recognition sequence. This mathematical approach has been developed into a computer program for the PC and is called AMINOMAT [3,7,9]. The program has been designed in order to allow the user to select (i) the target peptide length, (ii) the set of complementary amino acid to be assigned to each residue of the target sequence, (iii) the hydropathy scale to be used, and (iv) the averaging window r used for the score calculation. Target peptide length is usually selected between 5 and 20 residues. This is due to the fact that with more than 20 residues, the number of possible sequences to be considered usually exceed the program capacity. The averaging window used for calculations ranges from 7 to 11 residues, since preliminary studies have shown that peptides generated with these values display a higher affinity than peptides deduced with an averaging window of 1,3, or 5 residues. Since D, Q, N, and E are characterised by the same hydropathy index in the Kyte & Doolittle scale, the program considers them as a single complementary residue for calculation. Generally, given a defined target sequence, and a defined running length r, the program provides only one complementary sequence or a limited number of sequences with the same score and differing only for one or two residues. Sequences with higher scores are not considered and are not reported in the program printout.

Computer assisted identification of hydropathically complementary sequences.

Given the amino acid sequences of the two interacting polypeptides P_a and P_b, the corresponding moving average hydropathy values a_i and b_i are calculated in accordance with the formulae:

$$a_i = \Sigma\ h_k(P_a)/r$$

$$b_i = \Sigma\ h_k(P_b)/r$$

with S<i<n-S

where k is (i-S) to (i+S), r is an odd number defined as the run length (r<n) and S is equal to (r-1)/2 for each residue i of the first and second sequence. The hydropathic index h for each amino acid is obtained from the Kyte and Doolittle scale [12].

Subsequently, the hydropathic complementarity score, Θ_i defined in accordance with the formula:

$$\Theta_i = \Sigma\ \sqrt{(a_i+b_i)^2/L}$$

where L is the predefined searching length (3,5,7,9 or 11 residues), is computed for all the possible fragments of length L contained in P_a and P_b.

Preparation of multivalent complementary peptides.

Complementary peptides can be produced in multivalent forms by solid phase peptide synthesis following the Fmoc chemistry starting from a branched lysine core, obtained after subsequent couplings of Fmoc-Lys(Fmoc) on a Gly-HMP resin [6-9]. The central polylysine core results dimeric, tetrameric or octameric, according to the number of lysine coupling steps performed. In the case of poorly soluble complementary peptides, charged residues, such as arginine or glutamic acid, can be introduced in the core to improve solubility. In addition, polyglycine spacers, usually composed of three residues, can be introduced at the C-terminus of the complementary sequence to provide better accessibility for interaction after immobilisation on the solid surface. Purification from contaminants and reagents used in the cleavage procedure can be easily achieved by dialysis, since the resulting molecular weight is high enough to prohibit diffusion through the membrane. Multimeric complementary peptides can be chemically characterised by determination of amino acid composition after acid hydrolysis, or better, by determination of molecular weight by matrix-assisted laser desorption mass spectrometry.

Binding assay on microtiter plates.

Microtest II flexible assay plate (Falcon) were incubated with peptides (100 µl/well) in 0.1 M sodium carbonate buffer, pH 9.6, overnight at 4°C. After coating, plates were washed with PBS and 200 µl of PBS containing 3% BSA were then added to each well to block the uncoated plastic surface. The plates were then washed again with PBS containing 0.05% Tween 20 (PBS-T) and filled with samples containing mixtures of biotinylated peptides . For competition experiments, small amounts of the biotinylated material were mixed with competitors at different concentrations. After one hour incubation the plates were washed eight times with PBS-T, and wells were filled with 100 ml of a streptavidin-peroxidase solution diluted 1:100 with PBS-T containing 3% BSA (PBS-BT). After one hour, plates were washed, and development was carried out filling each well with a chromogenic substrate solution consisting of 1 mg/ml o-phenylendiamine in 0.1 M sodium citrate buffer, pH 5.0, containing 5 mM hydrogen peroxide. The color was allowed to develop for 30 minutes and blocked by adding 25 ml of 4.5 M sulphuric acid. The absorbance at 490 nm of each well was determined with a MIOS Microplate Reader (Merck).

RESULTS

Design, synthesis and characterisation of a complementary peptide to lipocortin fragment [16-32].

Human lipocortin 1 displays strong anti-inflammatory actions in rat and mouse [13]. A synthetic peptide corresponding to lipocortin residues 2-26 exhibits many biological effects similar to the native protein, such as the anti-inflammatory action, even if at higher concentrations [14]. The elucidation of the functional role of this portion of the lipocortin sequence would be facilitated by the availability of a specific ligand, to block or modulate lipocortin action as do neutralising monoclonal antibodies. For this reason, a peptide hydropathically complementary to the lipocortin sequence 16-32 [NEEQEYVQTVKSSKGGP], which includes the Tyr residue in position 22 previously shown as the key residue for biological action, has been designed using the AMINOMAT software. As expected, the complementary peptide [LIMKFGRLAALGGA] displayed a specular hydropathic profile in comparison with the lipocortin 16-32 target fragment. The complementary peptide has been then synthesised in a tetrameric form, starting from a tetradentate lysine core, to facilitate the characterisation of its binding properties for the corresponding target sequence (Figure 1).

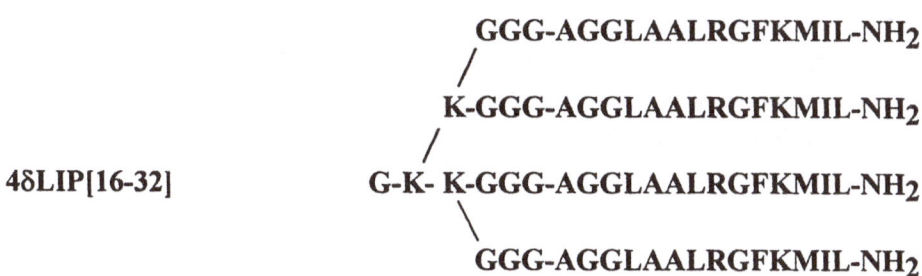

4δLIP[16-32]

GGG-AGGLAALRGFKMIL-NH$_2$

K-GGG-AGGLAALRGFKMIL-NH$_2$

G-K- K-GGG-AGGLAALRGFKMIL-NH$_2$

GGG-AGGLAALRGFKMIL-NH$_2$

Figure 1. Structure of 4δLIP[16-32], the complementary peptide to lipocortin fragment 16-32 synthesised in a tetrameric form. A triglycine spacer has been added at the C-terminus to provide augmented accessibility.

After derivatization with biotin, the lipocortin fragment recognised the tetrameric complementary peptide immobilised on to microtiter plates, in a dose dependent manner. Signal resulted also proportional to the amount of multimeric complementary peptide used in the coating procedure, while binding to plates derivatized with BSA was negligible (Figure 2).

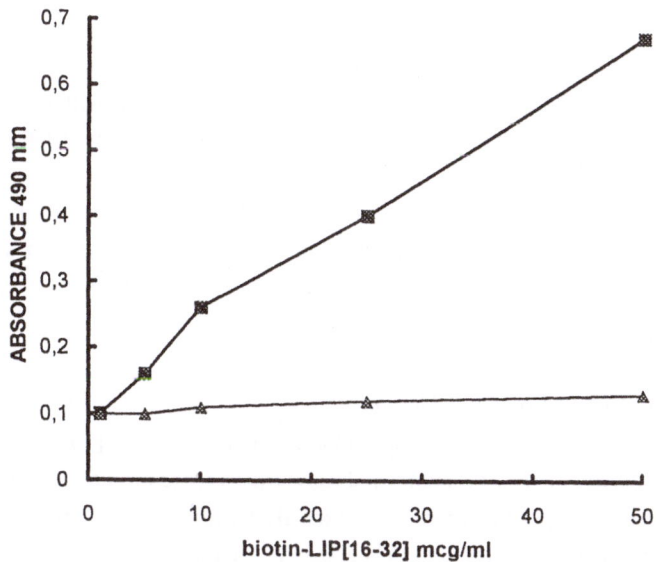

Figure 2. Binding of biotinylated LIP[16-32] to immobilised 4δLIP[16-32] at concentration 2.5 μg/well (squares), and to immobilised BSA at 10 μg/well (triangles).

Binding specificity was then further confirmed by the inhibitory effect of the tetrameric complementary peptide. As shown in Figure 3, presence of 4δLIP[16-32] in solution reduced the binding between biotinylated LIP[16-32] and the immobilised tetrameric complementary peptide in a dose dependent fashion. Inhibition was sequence specific, since an unrelated tetrameric peptide did not have any effect on the interaction.

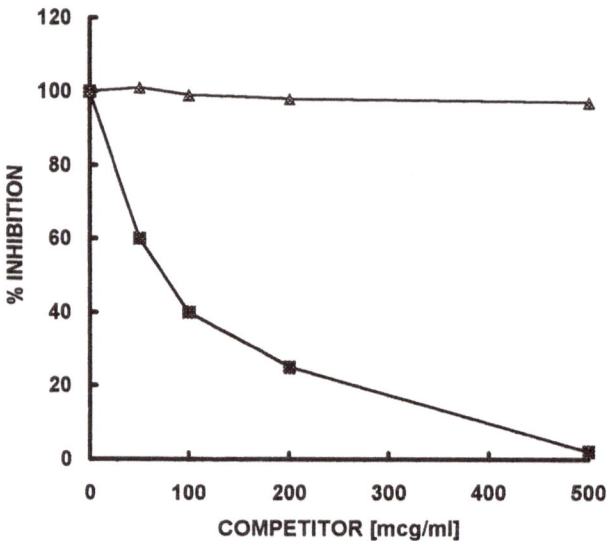

Figure 3: Competitive inhibition of biotin-LIP[16-32] to plates coated with 4δLIP[16-32] at 5 μg/well in the presence of 4δLIP[16-32] (squares) or unrelated peptide (triangles).

The multimeric peptide concentration yielding 50% inhibition was close to 1 μM. Results indicate clearly that the complementary peptide binds specifically to the target sequence, with an affinity sufficient for many applications. The biological activity of the complementary peptide is currently under investigation.

Identification of IL-1β/type I receptor binding site.

The IL-1 function is central to immune and inflammatory responses [15]. The mechanism regulating the multiplicity of IL-1 functions in different cells types are not clear but it is known that they are mediated by specific receptors. Two IL-1 receptors, IL-1RI and IL-1RII have been cloned and characterised [16,17]. They belong to the Ig superfamily of proteins as indicated by the structure of the extracellular binding regions, which are very similar in both proteins. In contrast the cytoplasmatic portion of the two receptors are different. IL-1RI contains approximately 219 amino acids in all species analysed so far, while human IL-1RII contains only 29 amino acids. Only the three-dimensional structure of the protein is known, while for the receptors only the amino acid sequence is available.

Starting from the assumption that at least part of the receptor binding site of IL-1β is a linear epitope, the ligand/receptor sequences have been examined using SITESEARCH [9], searching for sites hydropathically complementary.   Sequences with the highest level of hydropathic complementarity (score = 1.107) were identified in the mature IL-1β at position 88-99, and in the receptor sequence at position 152-163 of murine and 151-162 of human IL-1RI. The corresponding peptides were then produced by solid phase peptide synthesis and purified to homogeneity by RP-HPLC and their binding properties were evaluated by a solid phase binding assay [18].   The human or murine receptor fragments were immobilised on microtiter plates and treated with increasing amounts of IL-1β peptide derivatized with biotin . Complex formation was shown by incubation with avidin-peroxidase followed by chromogenic development.   Binding of biotin-IL-1β 88-99 peptide to microtiter plates coated with IL-1RI 151-162 peptide, was linearly dependent on the concentration of peptide added and on the amount of receptor fragment coated on to the plate, and saturable. Unrelated peptides, BSA, or scrambled IL-1β 88-99 peptide did not affect the reaction, while the unlabeled receptor fragment reduced the interaction to 50 % in concentration 3 μM   The murine fragment yielded very similar results, while whole IL-1β concentration yielding 50% inhibition was at least one order of magnitude lower than that observed with peptide fragments.   Accordingly, biotinylated IL-1β could bind to saturation and specifically to the IL-1R fragment, and binding was reduced in the presence of IL-1R fragment or IL-1β.   Also in this case the concentration of protein required to reduce to 50 % the interaction was lower than that required for the IL-1R receptor fragment. In order to verify that the IL-1β receptor fragment was able to recognise also full length IL-1β, experiments were performed on plates coated with IL-1β . As reported in Figure 4A, also in this case specific and saturable binding was observed between biotinylated receptor fragment 151-162 and IL-1β.   The biotinylated peptide had also a weak unspecific interaction with immobilised BSA, in a dose dependent fashion. Since all the binding and competition experiments were carried on in the presence of 5 mg/ml of BSA, this effect is probably due to the fact that after immobilisation on the plates, BSA undergoes denaturation processes, with exposure of sites responsible for the aspecific interaction.   Interaction between the two peptides was diminished in the presence of competing IL-1β[88-99] fragment (Figure 4B), with fifty percent inhibition reached with peptide concentration in the micromolar range. Unrelated peptide (scrambled IL-1β[88-99]) or BSA were unable to reduce the interaction. The competition effect of IL-1β [88-99] in the interaction indicates that the receptor fragment binds IL-1β through the [88-99] sequence.   Binding inhibition was observed also in the presence of unlabeled IL-1β receptor peptide.

The crystal structure of human IL-1β was determined independently in several laboratories [19].   The sequence of peptide 88-99, complementary to the 151-162 receptor fragment of human IL-1R, represents the long and highly exposed loop number 7 of IL-1β, with 6 arginine/lysine residues highly conserved in several species.

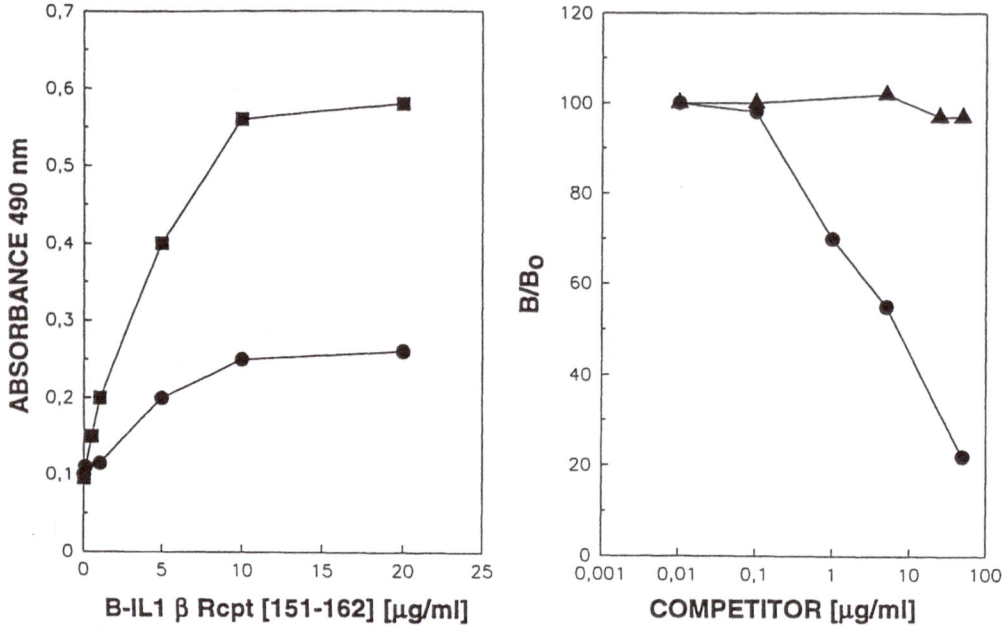

Figure 4. Left: Binding of human biotin-IL-1β[151-162] receptor fragment to microtiter plates coated with IL-1β (0.5 μg/well) [squares] and to BSA (10 μg/well) [circles]. Right: Competitive binding of human biotin-IL-1β[151-162] receptor fragment to plates coated with IL-1β (0.5 μg/well) in the presence of varying amounts of IL-1β [88-99] [circles], or unrelated peptide [triangles] (scrambled IL-1β fragment [88-99]). Reprinted from Ref. 18 with permission.

In order to test the validity of the hypothesis suggested by the computer analysis of the IL-1β/receptor sequences a mutational study of the 88-99 region (loop 7) of the protein was carried out [20].   As the entire loop sequence seemed to be involved in binding to the extracellular portion of the receptor, the entire loop, from position 89 to 97, was substituted. In order to attribute a loss of binding function to the mutant protein carrying the sequence substitution it was essential to ascertain the integrity of the core structure of the cytokine, which should be the same as that of wild type IL-1β. Two different sequences were used to replace loop 7, and only one gave a mutant with a structure similar to that of the wild type protein. The sequences GEGEESNDK and GTKGGQDIT were separately inserted in the loop 7 region of IL-1β replacing the amino acid sequence K88-R99 (loop 7/4 mutant) and N89-K97 (loop 7/12 mutant), respectively. The two sequences were derived from loop 4 and 12 of IL-1β.   Although both mutants were soluble and could be purified by affinity chromatography, loop 7/4 IL-1β showed a dramatic reduction of the receptor functions, suggesting structural alterations of the protein as compared with wild type IL-1β. Loop 7/12 mutant showed an affinity constant to IL-1RII identical to that of the wild type, while the binding to IL-1RI was decreased by more than 500 fold . This result showed that the substitution of the entire loop 7 with a sequence of the same length and different hydropathic profile generated a protein which could interact with unaltered affinity with the type II receptor, but was dramatically impaired in binding to IL-1RI. Since the binding of IL-1β to receptors I and II is independent , and the two IL-1β binding sites are different, results suggested that loop 7 is part of the binding site of IL-1β to IL-1RI, and that the amino acid sequence chosen for the substitution introduced a structural alteration limited to the IL-1β receptor binding region. The specificity of this result was confirmed by experiments in which the IL-1β sequence 48-56 (loop 4) was replaced by the sequence of loop 7 . The binding constant of this mutant for IL-1RI and II is identical to that of the wild type protein and the immunostimulatory and proinflammatory functions were also unaltered.

CONCLUSIONS

Design of complementary peptides to selected sequences representing exposed and accessible sequences of target polypeptides constitutes a simple approach to obtain in very limited time peptidyl ligands for the development of new lead compounds with potential therapeutic use. In addition, application of the MRT to protein/receptor interacting systems may provide useful informations regarding the precise localisation of recognition surfaces. Combined with site-directed mutagenesis, this approach may allow the modification of the biological properties of polypeptides involved in a wide variety of actions, and lead to redesigned molecules with new characteristics.

ACKNOWLEDGEMENTS

The author is grateful to Simona Germani, Menotti Ruvo, Antonio Verdoliva, and Matteo Villain for their contribution to various aspects of this work, and to Prof. Giovanni Cassani for his encouragement and support.

BIBLIOGRAPHY

1] Bost, K.L., Smith, E.M., and Blalock, J.E. (1985) Proc. Natl. Acad. Sci. USA 82,1372-1375.

2] Blalock, J.E. and Smith, E.M. (1984) Biochem. Biophys. Res. Comm. 121,203-207.

3] Fassina, G., Roller, P.P., Olson, A.D., Thorgeirsson, S.S., and Omichinski, J.G. (1989) . J. Biol. Chem. 264:11252-11257. .

4] Blalock, J.E. and Bost, K.L. (1986) Biochem. J. 234:679-683.

5] Shai, Y., Flashner, M., and Chaiken, I.M.; (1987) Biochemistry 26:669-675.

6] Tam, J.P. (1988) Proc. Natl. Acad. Sci. USA 85: 5409-5413.

7] Fassina, G., Corti, A., Cassani, G. (1992) Int. J. Pept. Prot. Res. 39:549-556.

8] Fassina, G. (1992) J. Chromatogr. 591,99-106.

9] Fassina, G., Cassani, G., and Corti, A. (1992) Arch. Biochem. Biophys. 296:137-143.

10]   Sloostra, J. W. and Roubos, E.W. (1991). In Antisense nucleic acids and proteins, fundamentals and applications. J.N.M. Mol and A.R. van der Krol (Eds.) pp.205-228 Marcel Dekker Inc., New York.

11] Fassina, G. (1989) Eur. Pat. Appl. EP 411,503.

12] Kyte, J and Doolittle, R.F. (1982) J. Mol. Biol. 157, 105-132.

13] Cirino, G., Flower, R.J., Browning, J.L., Sinclair, L.K., and Pepinsky, R.B. (1989) Proc. Natl. Acad. Sci. USA 86:3428-3432.

14] Cirino, G., Cicala, C., Sorrentino, L., Ciliberto, G., Arpaia, G., Perretti, M., and Flower, R.J. (1993) Br. J. Pharmacol. 108, 573-574.

15] Dinarello, C.A. (1989) Adv. Immunol. 44,153-205.

16] Sims, J.E., Acres, R.B., Grubin, C.E., McMahan, C.J., Wignall, J.M., March, C.J., and Dower, S.K. (1989) Proc. Natl. Acad. Sci. USA 86,8946-8950.

17] McMahan, C.J., Slack, J.L., Mosley, B., Cosman, D., Lupton, S.D., Brunton, L.L., Grubin, C.E., Wignall, J.M., Jenkins, N.A., Brannan, C.I., Copeland, N.G., Huebner, K., Croce, C.M., Cannizzaro, L.A., Benjamin, D., Dower, S.K., Spriggs, M.K., and Sims, M.J. (1991) Embo J. 10,2821-2832.

18] Fassina, G., Verdoliva, A., Cassani, G., and Melli, M. (1994) Growth Factors, 10, 99-106.

19] Priestle, J.P., Schaer, H.P., and Grutter, M.G. (1989) Proc. Natl. Acad. Sci. USA 86, 9667-9671.

20] Palla, E., Bensi, G., Solito, E., Tornese-Buonamassa, D., Fassina, G., Raugei, G., Spano, F., Galeotti, C., Mora, M., Domenighini, M., Rossini, M., Gallo, E., Carinci, V., Bugnoli, M., Bertini, F., Parente, L., and Melli, M. (1993) J. Biol. Chem 268,13486-13492.

AAS 46
Novel Molecular Approaches
to Anti-Inflammatory Theory
© 1995 Birkhäuser Verlag Basel

NITRIC OXIDE-RELEASING NSAIDs: A NOVEL CLASS OF GI-SPARING ANTI-INFLAMMATORY DRUGS

John L. Wallace[1], Quentin J. Pittman[2] and Giuseppe Cirino[3]

Departments of [1]Pharmacology & Therapeutics and [2]Medical Physiology, University of Calgary, Calgary, Alberta, Canada and [3]Department of Experimental Pharmacology, University of Naples, Naples, Italy

SUMMARY: The addition of a nitric oxide-releasing moiety to a number of common nonsteroidal anti-inflammatory drugs markedly reduces their toxicity in the gastrointestinal tract without interfering with their ability to inhibit prostaglandin synthesis. Moreover, the anti-inflammatory and anti-pyretic activities of the nitric-oxide releasing NSAID were comparable to the parent compound, while the anti-thrombotic activity *in vivo* was significantly enhanced. Nitric oxide-releasing NSAIDs may represent an alternative to existing anti-inflammatory, anti-pyretic and anti-thrombotic agents with greatly reduced toxicity in the gastrointestinal tract.

INTRODUCTION

Nonsteroidal anti-inflammatory drugs (NSAIDs) are among the most commonly used drugs. This widespread use is attributable to their broad spectrum of activities, which includes anti-inflammatory, anti-pyretic, anti-thrombotic and analgesic. However, the use of NSAIDs continues to be limited primarily by the ability of these compounds to produce damage in the gastrointestinal tract. While several strategies have been employed to reduce the ulcerogenic potential of NSAIDs, none have proven to be effective in reducing clinically significant adverse reactions (1). One of the problems associated with designing NSAIDs with reduced gastrointestinal toxicity is that the mechanism responsible for NSAID-induced gastroenteropathy remain poorly understood. The ability of NSAIDs to suppress prostaglandin synthesis appears to be critical in the pathogenesis of gastric ulceration induced by these agents, but is not of central importance in the production of small intestinal injury (2).

In recent years, a number of studies have supported a role for neutrophils in the pathogenesis of NSAID-induced gastric injury (3). This led us to suggest that interactions between neutrophils and the vascular endothelium are critical in the genesis of mucosal injury, and represent a potential target for gastroprotective drugs (3). NSAIDs may promote adhesive interactions between neutrophils and the endothelium by suppressing endothelial production of prostacyclin, a potent inhibitor of neutrophil activation and adherence. Another endothelial-derived inhibitor of neutrophil function is nitric oxide. We speculated that augmentation of endothelial nitric oxide production may counteract the detrimental effects of suppression of prostacyclin synthesis by NSAIDs, and therefore prevent the development of mucosal injury. In support of this, nitric oxide donors can reduce the severity of NSAID-induced gastric damage (4) while an inhibitor of nitric oxide synthase augmented NSAID-induced gastric damage (5). To further test this hypothesis, we investigated the possibility that linking a nitric oxide-releasing moiety to NSAIDs such as diclofenac would markedly reduce their ulcerogenic potential. In this paper, we describe the effects of nitrofenac (diclofenac nitroxybutylester) on the gastrointestinal tract and in models of acute inflammation, platelet aggregation and fever.

GASTROINTESTINAL DAMAGING PROPERTIES

The first system in which the NO-NSAID, nitrofenac, was compared to its parent NSAID, diclofenac, was in a model of acute gastric mucosal injury. Male, Wistar rats were fasted overnight then given diclofenac or nitrofenac at doses of 10-40 mg/kg orally. Five hours later, the rats were killed and the extent of hemorrhagic gastric damage was scored by an observer unaware of the treatments the rats had received. As shown in Figure 1, diclofenac caused extensive gastric damage which increased in severity in a dose-dependent manner. On the other hand, nitrofenac produced significantly less gastric damage at all doses tested, with only a few petechiae observed on each stomach. In order to determine if nitrofenac and diclofenac suppressed gastric prostaglandin synthesis to a similar extent, samples of the corpus region of the stomach were excised following the scoring of damage and incubated *in vitro* for determination of prostaglandin E_2 biosynthetic capacity (6). At each dose tested, nitrofenac suppressed gastric PGE_2 synthesis as effectively as diclofenac (7).

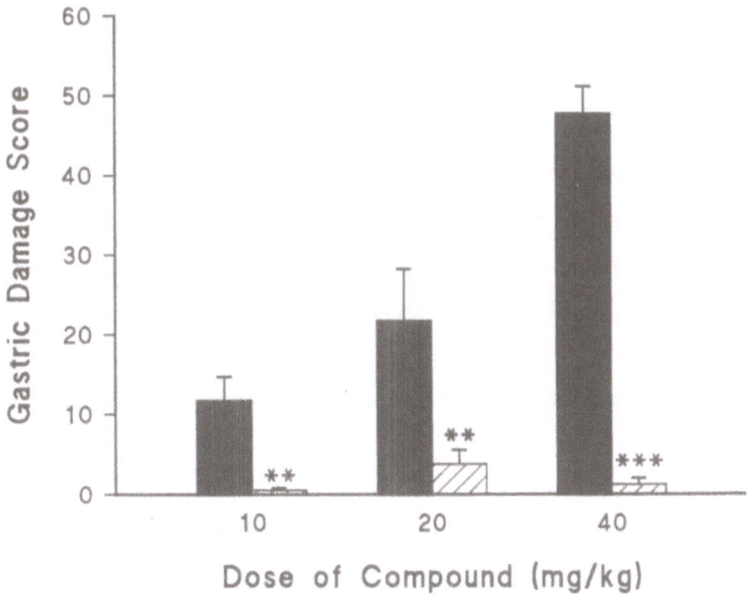

Figure 1: Gastric damage induced by diclofenac (solid bars) and nitrofenac (cross-hatched bars). Asterisks indicate significant differences between the two groups (**p<0.01, **p<0.001).

The damage produced in the stomach by diclofenac in the studies described above consisted of superficial erosions in the corpus region of the stomach. In man, the most significant gastric damage produced by NSAIDs is substantially deeper (penetrating into the submucosa) and occurs primarily in the antral region of the stomach. In order to assess the effects of nitrofenac in a model which more closely approximates human NSAID-induced ulcers, this compound was administered to rabbits twice-daily (subcutaneously) over several days. Diclofenac was given at a dose of 20 mg/kg, while nitrofenac was given at 30 mg/kg. The 30 mg/kg dose of nitrofenac contains an equivalent amount of diclofenac as a 20 mg/kg dose of the parent NSAID. The rabbits were killed 12 h after the 7th dose of the compounds and the stomach was examined. Penetrating ulcers in the antral region of the stomach were observed in 7 of the 9 rabbits treated with diclofenac, but in 0 of the 6 rabbits treated with diclofenac (7).

Repeated administration of NSAIDs to rats results in extensive small intestinal injury, ultimately leading to perforation and death. The mechanism responsible for the small intestinal injury induced by NSAIDs is not clear, but this damage does not appear to be attributable to cyclo-oxygenase suppression by the NSAIDs (2). Twice-daily subcutaneous administration of diclofenac to rats resulted in a progressive loss of weight and by day 8, all 10 of the rats receiving this treatment had died subsequent to intestinal perforation (8). On the other hand, rats treated with an equimolar dose of nitrofenac gained weight at a rate comparable to rats treated with the vehicle, and none of the rats died or exhibited small intestinal damage over a 14-day period of treatment. A similar lack of toxicity in the small intestine has been observed with other nitric oxide-releasing NSAID derivatives (9).

ANTI-INFLAMMATORY ACTIVITY

The anti-inflammatory activities of diclofenac and nitrofenac were compared in the carrageenan-induced paw edema model in rats. Groups of 5 rats each were given one of the two compounds at doses of 10-30 mg/kg, or the vehicle, orally. One hour later the rats were anesthetized with ether and carrageenan (0.1 ml) was injected into a hind foot pad. Paw volume was measured before carrageenan administration and hourly thereafter for 5 hours. Both diclofenac and nitrofenac dose-dependently reduced paw edema, with the maximal effect of the two drugs being about the same (7).

The anti-inflammatory effects of nitrofenac and diclofenac have also been compared in a model of chronic inflammation. Cuzzilin et al. reported that the two compounds were equally effective at reducing the severity of adjuvant arthritis in the rat (10).

ANTI-PYRETIC ACTIVITY

Fever was induced in conscious rats by intraperitoneal administration of endotoxin from *E. coli*, using a method similar to that described previously (11). Diclofenac (10 mg/kg), nitrofenac (15 mg/kg) or vehicle were administered intraperitoneally 2 hours after endotoxin administration and body temperature was measured by telemetry using intraperitoneally implanted thermistors for the following 10 hours. As shown in figure 2, administration of endotoxin resulted in a rapid increase in body temperature which was maintained throughout the duration of the experiment. Administration of diclofenac or nitrofenac two hours after endotoxin administration resulted in a significant attenuation of the pyretic response, with the responses to the two compounds being indistinguishable.

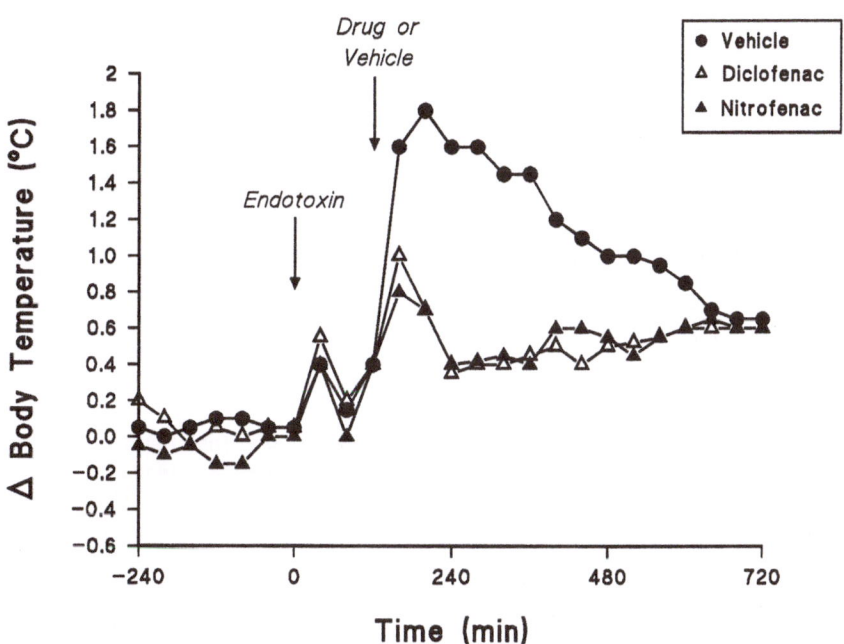

Figure 2: Effects of intraperitoneal diclofenac (10 mg/kg), nitrofenac (15 mg/kg) and vehicle on endotoxin-induced fever in rats. Each point represents the mean of 5 animals.

ANTI-THROMBOTIC ACTIVITY

The anti-thrombotic potential of nitrofenac was examined in both *in vitro* and *in vivo* systems. *In vitro* studies were performed using human platelets stimulated to aggregate by the addition of thrombin (6 units/ml). Platelets were preincubated with either diclofenac or nitrofenac for 10 minutes prior to being added into the cuvette of a platelet aggregometer. One minute later they were stimulated with thrombin and aggregation was monitored for the following 3 minutes. Both diclofenac and nitrofenac inhibited thrombin-induced platelet aggregation, with the concentration-response curves being virtually superimposable (Figure 3). This activity is almost certainly attributable to the ability of both compounds to suppress cyclo-oxygenase activity. Thus, no additional anti-platelet activity of nitrofenac which could be attributed to the release of nitric oxide was detectable in this *in vitro* system.

Inhibition of platelet aggregation *in vivo* was determined using the method of Pinon

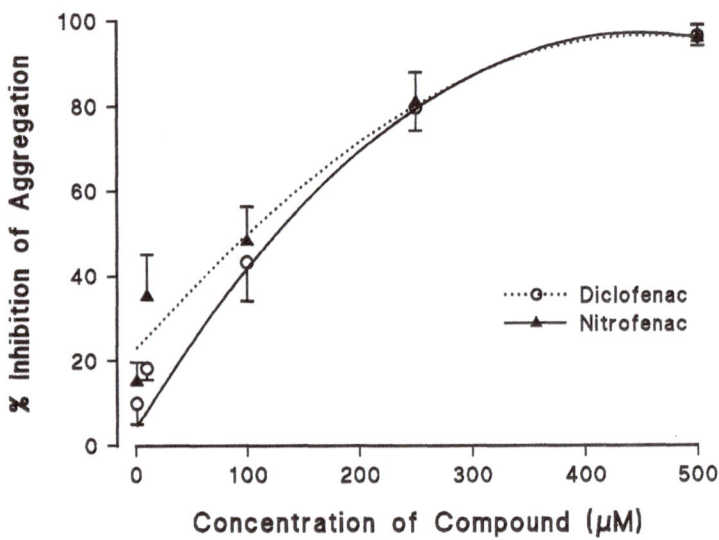

Figure 3:   Inhibitory effects of diclofenac and nitrofenac on thrombin-induced human platelet aggregation.   Each point represents the mean ± SEM of 5 separate platelet preparations.

(12).   Briefly, rats were orally pretreated with either diclofenac (1-10 mg/kg), equimolar doses of nitrofenac or the vehicle.   One hour later the rats were given collagen (2 mg/kg) intravenously, and three minutes thereafter, blood samples were taken.   The number of platelet aggregates was determined by light microscopy.   The ability of diclofenac and nitrofenac to inhibit platelet aggregation was determined by comparing the number of platelet aggregates to that seen in the vehicle-treated group. As shown in figure 4, collagen administration resulted in aggregation of approximately 45% of circulating platelets in the vehicle-pretreated group. Prior treatment with diclofenac dose-dependently reduced the number of platelet aggregates. Nitrofenac also dose-dependently reduced platelet aggregation, and at the higher doses, produced a significantly greater effect than equimolar doses of diclofenac. This observation suggests that nitrofenac has enhanced anti-platelet

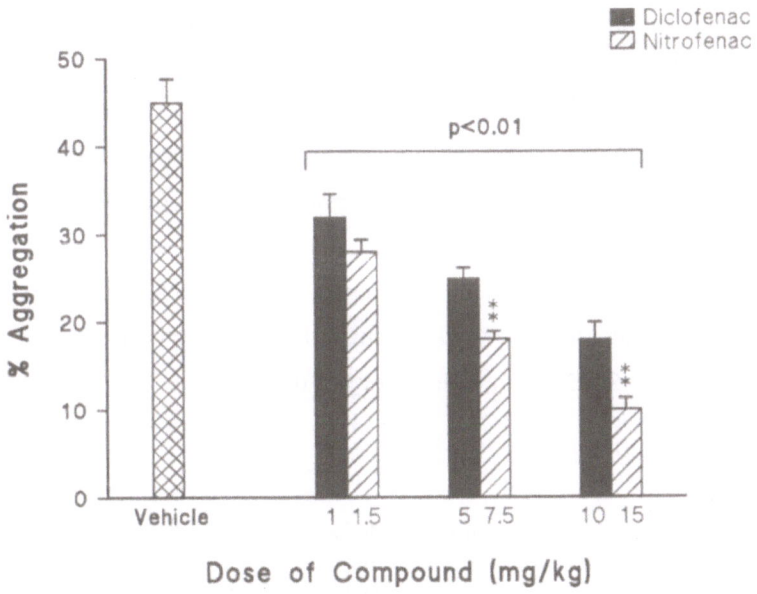

Figure 4: Inhibition of collagen-induced platelet aggregation *in vivo* in the rat by diclofenac and nitrofenac. Asterisks indicate a significant difference from the corresponding diclofenac group (**p<0.01). Each bar represents the mean ± SEM for 5 rats per group.

activity *in vivo*, a property that may be attributable to nitric oxide release from this compound. In support of this, we detected a significant increase in plasma nitrate/nitrite levels in rats treated with nitrofenac when compared to rats treated with diclofenac or vehicle (7).

SUMMARY

Nitrofenac is a nitric-oxide releasing derivative of diclofenac which has markedly less ulcerogenic activity, but comparable anti-inflammatory and anti-pyretic activity to the parent compound. While nitrofenac and diclofenac have similar anti-platelet activity *in vitro*, nitrofenac is significantly more effective at inhibiting platelet aggregation *in vivo*, presumably a consequence of the release of nitric oxide from this compound. While there is evidence to support the hypothesis that nitrofenac releases nitric oxide *in vivo*, the metabolism of this compound is not yet understood. Moreover, it is interesting that despite evidence that nitrofenac releases nitric oxide, it does not influence systemic blood pressure.

The results summarized in this paper suggest that nitrofenac and other nitric oxide-releasing NSAID derivatives may offer a useful alternative to existing NSAIDs. While the addition of a nitric oxide-releasing moiety to several NSAIDs greatly reduces their toxicity, it does not interfere with the ability of these compounds to inhibit prostaglandin synthesis, and therefore does not reduce their anti-inflammatory, anti-pyretic or anti-thrombotic activity. While the analgesic properties of nitrofenac and the other nitric oxide-releasing NSAIDs have not yet been systematically evaluated, the retained ability of these compounds to suppress cyclo-oxygenase activity would suggest that the compounds will exert similar activity to the parent NSAIDs from which they are derived.

REFERENCES

1. Soll AH, Weinstein WM, Kurata J, McCarthy D. Nonsteroidal anti-inflammatory drugs and peptic ulcer disease. Ann Intern Med 1991; 114: 307-319.
2. Whittle BJR. Temporal relationship between cyclooxygenase inhibition, as measured by prostacyclin biosynthesis, and the gastrointestinal damage induced by indomethacin in the rat. Gastroenterology 1981; 80:94-8.
3. Wallace JL. Gastric ulceration: Critical events at the neutrophil-endothelium interface. Can J Physiol Pharmacol 1993; 71:98-102.
4. Wallace JL, Reuter BK, Cirino G. Nitric oxide-releasing NSAIDs: a novel approach for reducing gastrointestinal toxicity. J Gastroenterol Hepatol 1994; in press.
5. Whittle BJR. Neuronal and endothelium-derived mediators in the modulation of the gastric microcirculation: integrity in the balance. Br J Pharmacol 1993; 110: 3-17.
6. Wallace JL, Keenan CM, Granger DN. Gastric ulceration induced by nonsteroidal anti-inflammatory drugs is a neutrophil-dependent process. Am J Physiol 1990; 259:G462-G467.
7. Wallace JL, Reuter B, Cicala C, McKnight W, Grisham MB, Cirino G. A diclofenac

derivative without ulcerogenic properties. *Eur. J. Pharmacol.* 1994; 257: 249-255.

8.   Reuter BK, Cirino G, Wallace JL. Markedly reduced intestinal toxicity of a diclofenac derivative. *Life Sci.* 1994; 55: PL1-PL8.

9.   Wallace JL, Reuter B, Cicala C, McKnight W, Grisham MB, Cirino G. Novel nonsteroidal anti-inflammatory drug derivatives with markedly reduced ulcerogenic properties in the rat. *Gastroenterology* 1994; 107: in press.

10.   Cuzzolin L, Conforti A, Donini M, Adami A, Del Soldato P, Benoni G. Effects of intestinal microflora, gastrointestinal tolerability and antiinflammatory efficacy of diclofenac and nitrofenac in adjuvant arthritic rats.   *Pharmacol. Res.* 1994; 29: 89-97

11.   Monda M, Pittman QJ. Cortical spreading depression blocks prostaglandin E_1 and endotoxin fever in rats. Am J Physiol 1993; 264: R456-R459.

12.   Pinon JF. In vivo study of platelet aggregation in the rat. J Pharmacol Methods 1984; 12: 79-84.

Novel Molecular Approaches
to Anti-Inflammatory Theory
© 1995 Birkhäuser Verlag Basel

ANTI-INFLAMMATORY LIPOCORTIN-DERIVED PEPTIDES

Mauro Perretti and Roderick J Flower

Department of Biochemical Pharmacology
The William Harvey Research Institute
Charterhouse Square
London EC1M 6BQ
United Kingdom

SUMMARY: Peptide Ac2-26, drawn from the sequence of human lipocortin 1, inhibited the release of elastase activity from cytoplasmic granules of human neutrophils, and neutrophil adhesion to monolayers of endothelial cells, in a concentration-dependent manner (approximate IC_{50} of 100 μg/ml, 33 μM). The effect of peptide Ac2-26 was not restricted to a specific neutrophil activator, being effective against formyl-Met-Leu-Phe (FMLP), leukotriene B_4 (LTB$_4$) and platelet-activating factor (PAF). Peptide Ac2-26 did not alter FMLP binding to its receptor. These *in vitro* observations complement *in vivo* data obtained with this peptide and may enable a better understanding of its pharmacology and, perhaps, that of lipocortin 1 too.

INTRODUCTION

Lipocortins (or annexins) are a family of calcium- and phospholipid-binding proteins which are still in search of a common biological function (1,2). Indeed, it is likely that different members of this family have different roles: for instance lipocortin 1 (LC1) possesses anti-inflammatory action (3) which appears to be different from that of another member like LC5 (4). A practical approach to a therapeutic application of the anti-inflammatory effect of lipocortin 1 (LC1) would be the identification of active segment(s) of its sequence (see Figure 1). The first successful attempt reported in the literature described a LC1 derived nonapeptide (antiflammin 2) with anti-inflammatory and anti-phospholipase A_2 properties (5): however, the efficacy of this peptide has not been fully confirmed and this has raised many doubts on the potential application of these nonapeptides (reviewed in (6)).

The crystal structure of LC5, and subsequently of a truncated form of LC1, have been elucidated and show that in both proteins the four conserved repeats refold to form a pore-like structure (2). In our investigation on LC1 mimetics, we have addressed our attention to the N-terminal region of LC1 sequence: this region is unique for each member of the family and, for instance, LC1 possesses a long N-terminus of 33 amino acids, whereas LC5 has only a truncated form of 6 amino acids. It is unfortunate that the orientation of this portion in LC1 structure is unknown, however it is likely to act as a modulator of the functions of the whole protein: original biological observations indicated that a LC1 isoform deprived of its inctact N-terminus was ineffective (6,7).

We have investigated the biological action of peptide Ac2-26 which corresponds to most of LC1 N-terminus, finding that it possesses inhibitory properties in the carrageenin-edema model (8). Moreover, this peptide mimicked LC1 in that it strongly reduced polymorphonuclear leukocyte (PMN) infiltration in a murine air-pouch model (9). From the latter study we inferred that PMN could represent a major target cell for the action of peptide Ac2-26. It is of interest that LC1 also dampens PMN activation (1), and that this cell type binds to LC1 in a specific and saturable fashion (10). LC1 binding sites are also present on murine PMN (11) and their occupation by an inactive LC1 preparation prevents peptide Ac2-26 action (9). In the present study we have investigated the effect of peptide Ac2-26 on PMN functions *in vitro*.

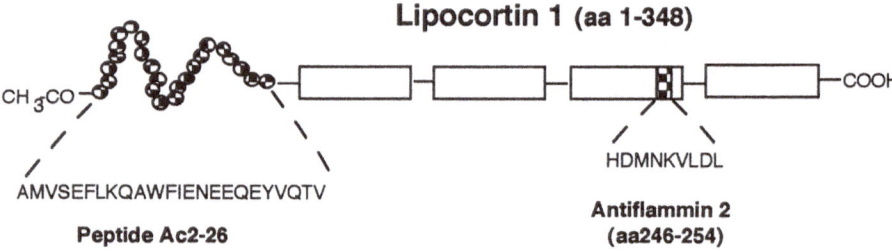

Figure 1. Here is depicted the lipocortin 1 structure with the four similar repeats (open boxes) and the N-terminus (circles). Peptides reported to be biologically active are shown: peptide Ac2-26 corresponding to most of the N-terminus region, and antiflammin 2 corresponding to aa 246-254 on the third repeat.

MATERIALS AND METHODS

Peptide Ac2-26 derived either from human LC1 (acetyl-A-M-V-S-E-F-L-K-Q-A-W-F-I-E-N-E-E-Q-E-Y-V-Q-T-V-K), murine LC1 (acetyl-A-M-V-S-E-F-L-K-Q-A-R-F-L-E-N-Q-E-Q-E-Y-V-Q-A-V-K) or rat LC1 (acetyl-A-M-V-S-E-F-L-K-Q-A-C-Y-I-E-K-Q-E-Q-E-Y-V-Q-A-V-K) sequences were generously synthetised by Dr M Toda (ONO Pharmaceutical Co., Osaka, Japan). Peptide preparations were more than 95% pure as analysed by HLPC. Amino acid composition and Mr were confirmed by mass spectrometry. Solutions were made fresh in sterile PBS supplemented with 0.2% low endotoxin BSA .

Circulating PMN were prepared from citrated blood collected from healthy volunteers. Blood, diluted 1:2 in sterile PBS, was layered over a histopaque (1077/1119) gradient and centrifuged at 400 g for 30 min at room temperature. The layer containing PMN (bottom) was then collected, the cells were washed twice in PBS and red blood cell removed by hypotonic lysis.

PMN adhesion to Endothelial Cells

Adhesion between the endothelial cell line, EA.hy926 (a generous gift of Dr Cora-Jean S Edgell) (12) and freshly prepared human neutrophils was assessed in these experiments. The endothelial cells were grown in DMEM containing 10 % foetal calf serum in T-75 flasks and subcultured 1:3 approximately once a week. For the adhesion assay 2 x 10^4 cells per well were seeded in 96-well plates and used 3 days later when homogenous monolayers had formed. Neutrophils were then suspended in the adhesion assay medium HBSS + bovine serum albumin (0.2 % w/v) + Ca^{2+} and Mg^{2+} (1.3 mM) to give 10^5 cells in 100μl total volume and added to endothelial monolayers for an incubation period of 30 min at 37°C. Peptides were incubated with PMN 5 min prior to FMLP addition. The chemoattractant was used at a final concentration of 0.1 μM, chosen on the basis of preliminary experiments. At the end of the incubation period, non-adherent cells were washed off, and the adhesion measured by assessment of myeloperoxidase actvity. To do this, cells were dissolved in Triton X100 1% (100 μl) followed by addition of both hydrogen peroxide 0.5 % (25 μl) and ortho-dianisidine 4.7 mM (25 μl) and left in contact for 30 min. The reaction at the end of this time was arrested with sodium azide 0.4 % and the absorbance read at 450 nm in an ELISA plate reader (Anthos Labtec). The % adhesion was then calculated in comparison to the absorbance measured in samples containing the total number of neutrophils added in each well. Values are reported as % inhibition calculated on the incidence of adherence in the absence of peptide addition. Usually ≈20% unstimulated adhesion was measured which rose to ≈ 65% when PMN were stimulated with FMLP.

Release of Elastase from Stimulated PMN

The assay was performed as recently described (13). Briefly, 2-4 x10^6 PMN, prepared as described above, were incubated in a water bath at 37° C in 0.4 ml of RPMI 1640 + 0.1% BSA containing cytochalasin B (5 μg/ml), in the absence or presence of different concentrations of peptide Ac2-26 for 5 min. Then different stimuli were added, in 0.1 ml volumes at concentrations chosen on the basis of preliminary experiments, and the incubation carried out for further 15 min. Cell stimulation was stopped on ice, and the supernatants collected following centrifugation in a micro-centrifuge at 6,500 rpm for 1 min. Elastase activity in the supernatants was measured as nmol of p-nitroaniline (*p*-NA) released from the substrate methoxy-succinyl-A-A-P-V-*p*-NA (100mM; Sigma) following 30 min incubation in 96-well plates. The reaction was stopped by addition of acetic acid (50%) and plates were read at 405 nm in an ELISA plate reader (Anthos Labtec). Data are reported as % inhibition calculated on the enzymatic activity ($\approx$ 10-15 nmol/min/10^6 PMN vs. an unstimulated release of 0.5-1.0 nmol/min/million cells) measured in the absence of peptide.

FMLP-Binding to Human PMN

To evaluate peptide Ac2-26 effect upon FMLP binding, 0.5-1.0 x10^6 PMN were suspended in 20 μl HBSS + BSA 0.2% and incubated on ice in a 96-well plate for 5 min with or without peptide Ac2-26 prior to the addition of FITC-conjugated FMLPL (Peninsula Laboratories) (14). After 1 h the cells were washed twice with HBSS + 0.2% BSA (200 μl). Cells were then fixed in an equal volume of 2% paraformaldhyde in PBS. Flow cytometry analysis was performed using a FACScan II analyser (Becton Dickinson) with air-cooled 100 mW argon ion laser tuned to 488 nm and Consort 32 computer running Lysis II software. Data are reported as units of fluorescence (mean fluorescence intensity, MFI, in FL-1 channel).

RESULTS

Peptide Ac2-26 did not interfere with FMLP binding as assessed with the FMLPL-FITC probe. In a representative experiment, MFI values of 83 and 61 were measured at the agonist concentrations of 1 and 0.1 μM, respectively, these figures being 96 and 60 when cells were preincubated with 100 μg/ml of peptide Ac2-26.

However, stimulation of PMN adhesion to the endothelial monolayers with FMLP was significantly attenuated in the presence of peptide Ac2-26 (Figure 2, top panel). Preincubation of cells with human Ac2-26 resulted in a concentration-dependent inhibition of PMN adhesion with an approximate IC$_{50}$ of 100 μg/ml. Murine and rat Ac2-26 were considerably less active in inhibiting cell adhesion: at 100 μg/ml they gave only 16 $\pm$ 9 and 13 $\pm$ 9 of inhibition (mean $\pm$ SEM of 3 experiments performed in triplicate), respectively.

Elastase release from activated PMN was also inhibited by human peptide Ac2-26 in a concentration-dependent fashion (Figure 2, bottom panel) again with an approximate IC$_{50}$ of 100 μg/ml when calculated against FMLP (data from 3 experiments). In this case the effect

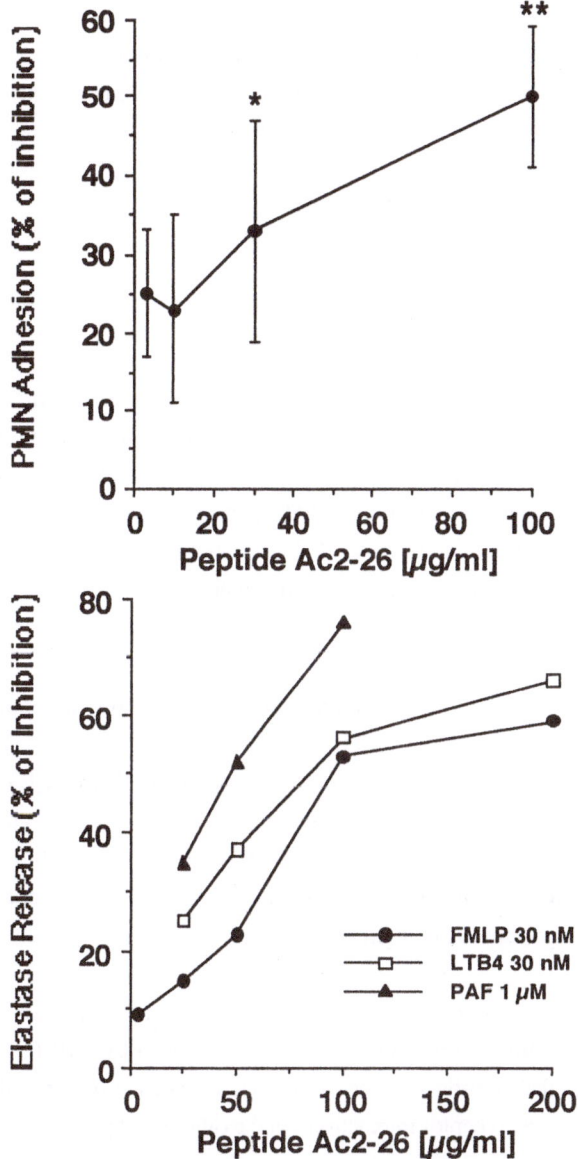

Figure 2. Top panel: Peptide Ac2-26 inhibits FMLP (0.1 μM)- induced PMN adhesion to endothelial monolayers. Data are mean ± SEM of 4 experiments performed in triplicate. * and ** $p < 0.05$ and 0.01 vs control (as calculated on original values). Bottom panel: Peptide Ac2-26 inhibits the release of elastase from stimulated human PMN. SEM are omitted for reason of clarity. Inhibitions with concentrations of 100 and 200 μg/ml are significant.

of peptide Ac2-26 was also assessed on different stimuli. Figure 2 (bottom panel) shows that a significant reduction of PAF (1 experiment) or LTB$_4$ (3 experiments) induced elastase release was obtained when PMN were preincubated with peptide Ac2-26. Finally, the potential interference of this peptide with the elastase/substrate interaction was ruled out on the basis of 3 experiments were peptide Ac2-26 was added to the cell supernatant after the stimulation had occured: no changes in enzymatic activity were then observed (not shown).

DISCUSSION

This study focuses upon the pharmacological effects of a peptide corresponding to the N-terminus region of LC1 sequence. Peptide Ac2-26 has already been shown to inhibit PMN migration into an inflammatory site in response to local application of interleukin-1 or interleukin-8 (9). Moreover, treatment of animals with peptide Ac2-26 inhibited PMN-dependent edema in mouse skin (9). The process of the leukocyte extravasation has been recently been partially elucidated: an interaction between the neutrophil and the endothelium at the level of the post-capillary venules is the initial step (15). On the basis of all these *in vivo* models we could not identify the target cell which was affected by treatment with the peptide, i.e. the neutrophil or the endothelial cell. Although it appears that circulating neutrophils are a target for the parent protein (see introduction), this being reinforced by the existence of specific and saturable binding sites (11), a direct action of peptide Ac2-26 on this cell type awaits definite confirmation. We have therefore set up a series of models *in vitro* to clarify this aspect. Incubation of human neutrophils with peptide Ac2-26 resulted in an inhibition of cell activation assessed both as release of a proteolytic enzymatic activity and as adhesion to endothelial monolayers. It is important to highlight that:

i) the effect of peptide Ac2-26 was not dependent upon the stimulus applied, as it produced a significant ($\approx 50\%$) inhibition on elastase release provoked by FMLP, LTB$_4$ and PAF; this suggests that the peptide is affecting a specific activation step of the neutrophil downstream the agonist/receptor interaction; on the other hand peptide Ac2-26 did not appear to interact with the FMLP receptor as assessed with the fluorocytometric technique.

ii) the peptide Ac2-26 was an effective inhibitor of human PMN functions in the adhesion assay; however, both the rat and murine LC1-derived peptides showed a much less efficacy, being essentially inactive upon human PMN: this clearly indicates the existence of a one-way species specificity, with the human peptide Ac2-26 being active on human and murine cells (9) and the murine and rat peptides being inactive on human PMN but effective on murine cells (unpublished data). Such a phenomenon has been already described for other proteins (e.g. interleukin-10) and must be taken into account for future evaluation of the biological efficacy of LC1 and LC1-derived peptides when obtained from different species.

iii) in the two *in vitro* experimental systems peptide Ac2-26 inhibited cell activation with similar potency, with an approximate IC_{50} of 100 μg/ml (33 μM) in both assays. This again points to a similar cellular mechanism(s) affected by incubation with the peptide Ac2-26.

In conclusion, in this study we have broadened the spectrum of biological activities ascribed to a LC1 peptide derived from the unique N-terminus region of this protein, adding to the *in vivo* evidence a series of *in vitro* observations which complement the pharmacology of peptide Ac2-26: the circulating neutrophil represents an important target cell for this peptide and, probably, for the parent protein itself.

ACKNOWLEDGMENTS

This work was supported by ONO Pharmaceutical Co. (Osaka, Japan). We thank Miss SK Wheller for technical assistance.

REFERENCES

1) Flower RJ and Rothwell NJ. Lipocortin-1: cellular mechanisms and clinical relevance. Trends Pharmacol Sci 1994; 15: 71-76.

2) Raynal P and Pollard HB. Annexins: the problem of assessing the biological role for a gene family of multifunctional calcium- and phospholipid-binding proteins. Biochim Biophys Acta 1994; 1197: 63-93.

3) Cirino G, Peers SH, Flower RJ, Browning JL and Pepinsky RB. Human recombinant lipocortin 1 has acute local anti-inflammatory properties in the rat paw edema test. Proc Natl Acad Sci USA 1989; 86: 3428-3432.

4) Becherucci C, Perretti M, Solito E, Galeotti C and Parente L. Conceivable difference in the anti-inflammatory mechanisms of lipocortin 1 and 5. Med Inflamm 1993; 2: 109-113.

5) Miele L, Cordella-Miele E, Facchiano A and Mukherjee AB. Novel anti-inflammatory peptides from the region of highest similarity between uteroglobin and lipocortin I. Nature 1988; 335: 726-730.

6) Perretti M. Lipocortin-derived peptides. Biochem Pharmacol 1994; 47: 931-938.

7) Browning JL, Ward MP, Wallner BP and Pepinsky RB. Studies on the structural properties of lipocortin-1 and the regulation of its synthesis by steroids. In Cytokines and lipocortins in inflammation and differentiation. (Eds. M. Melli and L. Parente) pp. 27-45, Wiley-Liss, New York 1990.

8) Cirino G, Cicala C, Sorrentino L, Ciliberto G, Arpaia A, Perretti M and Flower RJ. Anti-inflammatory actions of an N-terminal peptide from human lipocortin 1. Br J Pharmacol 1993; 108: 573-574.

9) Perretti M, Ahluwalia A, Harris JG, Goulding NJ and Flower RJ. Lipocortin-1 fragments inhibit neutrophil accumulation and neutrophil-dependent edema in the mouse: a qualitative comparison with an anti-CD11b monoclonal antibody. J Immunol 1993; 151: 4306-4314.

10) Goulding NJ, Luying P and Guyre PM. Characteristics of lipocortin 1 binding to the surface of human peripheral blood leucocytes. Biochem Soc Trans 1990; 18: 1237-1238.

11) Perretti M, Flower RJ and Goulding NJ. The ability of murine leukocytes to bind lipocortin 1 is lost during acute inflammation. Biochem Biophys Res Comm 1993; 192: 345-350.

12) Edgell C-J, McDonald CC and Graham JB. Permanent cell line expressing human factor VIII-related antigen established by hybridization. Proc Natl Acad Sci USA 1983; 50: 3734-3737.

13) Iwamura H, Moore A and Willoughby DA. Interaction between neutrophil-derived elastase and reactive oxygen species in cartilage degradation. Biochim Biophys Acta 1993; 1156: 295-301.

14) Leonard EJ, Noer K and Skeel A. Analysis of human monocyte chemoattractant binding by flow cytometry. J Leukoc Biol 1985; 38: 403-413.

15) Springer TA. Traffic signals for lymphocyte recirculation and leukocyte emigration: the multistep paradigm. Cell 1994; 76: 301-314.

AAS 46
Novel Molecular Approaches
to Anti-Inflammatory Theory
© 1995 Birkhäuser Verlag Basel

INHIBITION OF LIPID MEDIATOR BIOSYNTHESIS IN HUMAN

INFLAMMATORY CELLS BY BIRM 270

Thomas P. Parks[1], Ann F. Hoffman[1], Carol A. Homon[1], Anne G. Graham[1], Edward S.Lazer[1], Floyd H. Chilton[2], Pierre Borgeat[3], Donald Raible[4], Edward Schulman[4], David A. Bass[2], and Peter R. Farina[1]

[1]Boehringer Ingelheim Pharmaceuticals, Inc., Ridgefield, CT 06877, U.S.A., [2]Bowman Gray School of Medicine, Winston-Salem, NC 27157-1054, U.S.A., [3]Centre de recherche du CHUL et Universite Laval, Sainte-Foy, Québec G1V 4G2, Canada, [4]Hahnemann University, Philadelphia, PA 19102, U.S.A.

Summary

BIRM 270 was developed as a potent and enantioselective inhibitor of LTB_4 biosynthesis by human neutrophils, and was also found to inhibit LTC_4 production by human eosinophils and lung mast cells. BIRM 270 inhibited LTB_4 synthesis in neutrophils by preventing arachidonate release from membrane phospholipids, and over the same concentration range, inhibited PAF biosynthesis. BIRM 270 did not directly inhibit acylhydrolases which have been implicated in eicosanoid and PAF biosynthesis, suggesting an indirect mode of action.

Discovery and Cellular Effects

Leukotrienes (LTs) are lipid-derived mediators which have been implicated in the pathology of numerous immunologic and inflammatory disorders, including asthma, allergic rhinitis, rheumatoid arthritis, inflammatory bowel disease, and psoriasis [1]. For this reason, the regulation of LT biosynthesis has received considerable attention as a target for drug development. Several years ago we began a search for novel LT biosynthesis inhibitors using ionophore A23187-stimulated LTB_4 release from human neutrophils as a screen. During random screening of a synthetic chemical library, a compound was identified which inhibited

LTB$_4$ biosynthesis with an IC$_{50}$ of 230 nM. Other than inhibition of LT production, the compound had little effect on other neutrophil functions, was not cytotoxic, and was not active in a general receptor screen. These intriguing characteristics, and a vague structural resemblance to the Merck 5-lipoxygenase activating protein (FLAP) inhibitor, MK886 [2], prompted us to launch a synthetic program to optimize the LT inhibitory activity. Following extensive chemical modification, BIRM 270 emerged with an IC$_{50}$ of 1 nM [3]. As shown in Figure 1, BIRM 270 possessed a chiral center, and the (S)-enantiomer was about 40-fold more potent than the (R)-enantiomer BIRM 271 [4].

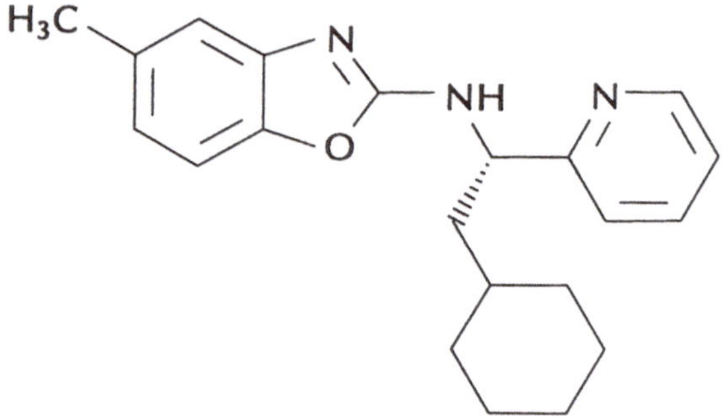

Figure 1. Structure of BIRM 270

Although BIRM 270 potently inhibited LTB$_4$ production induced by ionophore, we were concerned whether the compound could inhibit receptor-stimulated LT biosynthesis. Without priming, most receptor-mediated stimuli produce little if any LTB$_4$ in human neutrophils. Therefore, we employed an experimental procedure which greatly potentiated the ability of a weak stimulus, platelet-activating factor (PAF), to elicit LTB$_4$ production. A brief priming treatment with the chemotactic peptide, fMetLeuPhe, which did not itself stimulate LTB$_4$ synthesis, considerably augmented the subsequent effects of a PAF challenge. BIRM 270 inhibited LTB$_4$ biosynthesis elicited in this manner with an IC$_{50}$ of about 5 nM.

The inhibition of LT biosynthesis by BIRM 270 was not restricted to human neutrophils. The compound also inhibited LTC$_4$ production by ionophore-stimulated eosinophils, isolated from the blood of atopic donors, and human lung mast cells stimulated with either ionophore or IgE receptor crosslinking (summarized in Figure 2). Half-maximal inhibition of these reponses by BIRM 270 was typically in the 15-30 nM range. As expected, the (R)-enantiomer, BIRM 271, was less potent than BIRM 270.

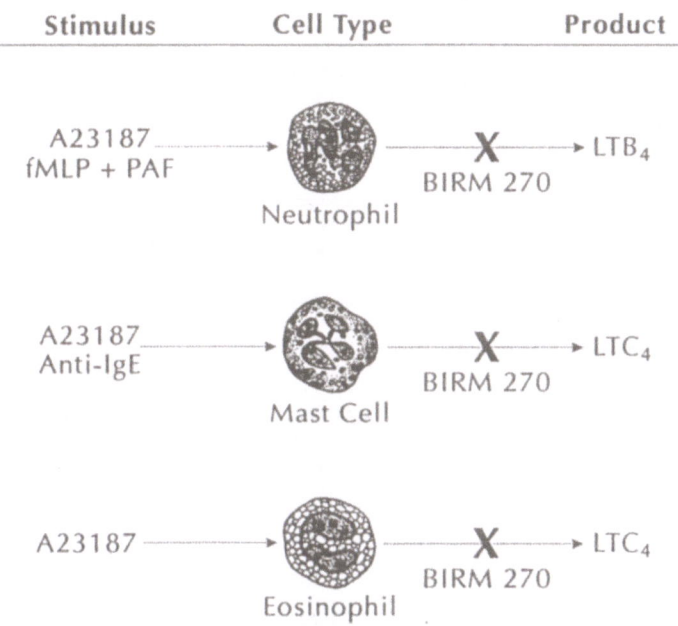

Figure 2. Effects of BIRM 270 on ionophore A23187- and receptor-stimulated leukotriene biosynthesis by human inflammatory cells.

Mechanism of Action

HPLC analysis of arachidonate metabolites from ionophore-stimulated neutrophils revealed that BIRM 270 suppressed the formation of all 5-LO reaction products, including 5-HETE, LTB_4 and its ω-oxidation products, and other LTA_4-derived products. This observation initially suggested that the compound might have inhibited 5-lipoxygenase (5-LO) or its activator protein, FLAP [5]. However, BIRM 270 did not appear to directly inhibit 5-LO enzymatic activity in a cell-free assay, nor did it compete with [^{3}H]MK886, a FLAP inhibitor [2], in a FLAP binding assay using neutrophil membranes (summarized in Figure 3).

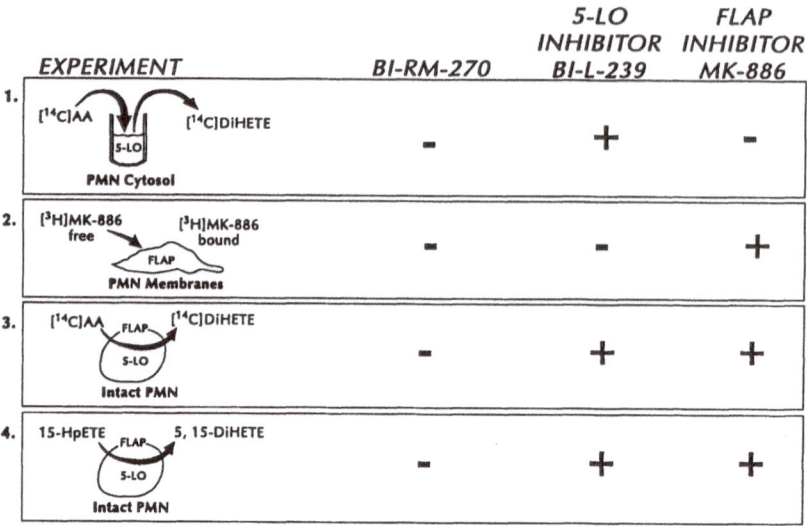

EXPERIMENT	BI-RM-270	5-LO INHIBITOR BI-L-239	FLAP INHIBITOR MK-886
1. [^{14}C]AA → [^{14}C]DiHETE (5-LO, PMN Cytosol)	–	+	–
2. [^{3}H]MK-886 free → [^{3}H]MK-886 bound (FLAP, PMN Membranes)	–	–	+
3. [^{14}C]AA → [^{14}C]DiHETE (FLAP, 5-LO, Intact PMN)	–	+	+
4. 15-HpETE → 5,15-DiHETE (FLAP, 5-LO, Intact PMN)	–	+	+

Figure 3. Summary of the evidence indicating that BIRM 270 does not directly inhibit 5-lipoxygenase or FLAP. Inhibitory activity is denoted +.

In activated intact neutrophils, the 5-LO/FLAP system can convert exogenous substrates, such as [^{14}C]arachidonate and 15-hydroperoxy-arachidonate (15-HpETE), to 5-oxygenated products, [^{14}C]diHETE (i.e. LTB$_4$) and 5,15-diHETE, respectively [6]. Unlike direct inhibitors of 5-LO (i.e. BI-L-239 [7]) and FLAP (i.e. MK886 [2]), BIRM 270 did not block the metabolism of exogenous 5-LO substrates (summarized in Figure 3). Although the conversion of exogenous 15-HpETE to 5,15-diHETE was unaffected by BIRM 270, the compound potently inhibited production of 5-HETE and LTB$_4$ from endogenous arachidonate in the same cells [4]. These results suggested that BIRM 270 acted at the level of substrate availability, rather than downstream metabolism like other well-known inhibitors of leukotriene biosynthesis. Consistent with this concept, the blockade of LTB$_4$ biosynthesis by BIRM 270 was overcome by the addition of exogenous arachidonate [4].

To confirm that BIRM 270 inhibited LTB$_4$ biosynthesis at the level of substrate availability, we measured endogenous free arachidonate levels in neutrophils by gas chromatography/mass spectrometry. BIRM 270 blocked ionophore-stimulated arachidonate release at 30-100 nM, with half-maximal inhibition at about 14 nM. BIRM 271, the (R)-enantiomer, was approximately 10-fold less potent. BIRM 270 did not appear to affect the reacylation of arachidonate into phospholipids, indicating that the compound specifically affected the deacylation pathway. Under identical experimental conditions, the dose-response curves for inhibition of arachidonate release and LTB$_4$ biosynthesis were superimposable, indicating that the blockade of arachidonate release by BIRM 270 could fully account for the inhibition of LTB$_4$ production.

Ether-linked phospholipids represent the largest source of arachidonate mobilized in ionophore-stimulated human neutrophils [8]. Since 1-*O*-alkyl-2-arachidonoyl-*sn*-glycero-3-phosphocholine also serves as a precursor for PAF, it has been proposed that arachidonate release and PAF biosynthesis may be coordinately regulated [9]. We used BIRM 270 as a tool to test this hypothesis, and found that the compound inhibited PAF biosynthesis by ionophore-stimulated neutrophils over the same concentration range as inhibition of LTB$_4$ biosynthesis and arachidonate release (depicted in Figure 4) [4]. In a parallel experiment, the production of lyso-PAF was similarly inhibited by BIRM 270.

Figure 4. Summary of the effects of BIRM 270 on ionophore A23187-stimulated lipid mediator biosynthesis by human neutrophils.

Since BIRM 270 inhibited the deacylation of arachidonate-containing phospholipids in stimulated neutrophils, it followed that the compound must have compromised the activity of an *sn*-2 acylhydrolase (e.g. PLA$_2$ shown in Figure 4). Neutrophils contain at least three enzymes which have been implicated in arachidonate mobilization and/or PAF biosynthesis: an 85 kDa intracellular PLA$_2$ [10,11], a 14 kDa type II secretory PLA$_2$ [12], and an arachidonoyl-selective, CoA-independent transacylase [13-15]. BIRM 270 did not directly inhibit the activity of any of these enzymes in cell-free assays, suggesting an indirect mechanism of inhibition. The activity of the intracellular 85 kDa PLA$_2$ may be regulated by Ca^{2+}-dependent membrane association [16,17], and by phosphorylation [18,19]. We are currently investigating whether BIRM 270 interferes with either of these two processes in intact neutrophils. Preliminary evidence suggested that the compound did not affect association of the recombinant 85 kDa enzyme with neutrophil membranes.

The non-pancreatic 14 kDa secretory PLA_2 has generally been viewed as a pro-inflammatory mediator and component of the neutrophil anti-microbial arsenal [20,21]; its role in arachidonate mobilization has been unclear. However, recent studies using murine bone marrow-derived mast cells have suggested that the regulated secretion of this enzyme may be coupled to arachidonate release and eicosanoid biosynthesis [22]. Human neutrophils stimulated with ionophore A23187 have also been shown to release PLA_2 activity into the surrounding medium [23]. This activity was DTT-sensitive and neutralized by treatment with specific antibodies directed against the 14 kDa human group II secretory PLA_2 . Preliminary evidence indicated that the extracellular appearance of the PLA_2 activity could be effectively blocked by pre-treating neutrophils with BIRM 270. Since the compound did not inhibit PLA_2 activity of the supernatant fluid, it would appear that BIRM 270 blocked secretion rather than catalytic activity. These results suggest the intriguing possibility that the blockade of PLA_2 release by BIRM 270 may be related to the inhibition of arachidonate release and PAF biosynthesis. Studies are in progress to further define the role of secretory PLA_2 in arachidonate mobilization, and to determine how BIRM 270 impacts on this process.

Acknowledgements

The authors gratefully acknowledge the secretarial and graphics assistance of Ms. Janet Abbott and Ms. Leigh Rondano, and the considerable intellectual and technical contributions made by many additional individuals.

References

1. Lewis RA, Austen KF, Soberman RJ. Leukotrienes and other products of the 5-lipoxygenase pathway. Biochemistry and relation to pathobiology in human disease. N Engl J Med 1990; 323:645-655.

2. Gillard J, Ford-Hutchinson AW, Chan C, Charleson C, Denis D, Foster A, Fortin R, Leger S, McFarlane CS, Morton S, Piechuta H, Riendeau D, Rouzer CA, Rokach J, Young R, MacIntyre DE, Peterson L, Bach T, Eirmann G, Hopple S, Humes J, Hupe L, Luell S, Metzger J, Meurer R, Miller DK, Opas E, Pacholok S. L-663,536 (MK-886) (3-[1-(4-chlorobenzyl)-3-*t*-butyl-thio-5-isopropylindol-2-yl]-2,2-dimethylpropanoic acid), a novel, orally active leukotriene biosynthesis inhibitor. Can J Physiol Pharmacol 1989; 67:456-464.

3. Lazer ES, Miao CK, Wong HC, Sorcek R, Spero DM, Gilman A, Pal K, Behnke M, Graham AG, Watrous JM, Homon CA, Nagel J, Shah A, Guindon Y, Farina PR, Adams J. Benzoxazolamines and benzothiazolamines: potent, enantioselective inhibitors of leukotriene biosynthesis with a novel mechanism of action. J Med Chem 1994; 37:913-923.

4. Farina PR, Graham AG, Hoffman AF, Watrous JM, Borgeat P, Nadeau M, Hansen G, Dinallo RM, Adams J, Miao CK, Lazer ES, Parks TP, Homon CA. BIRM 270: A novel inhibitor of arachidonate release which blocks leukotriene B$_4$ and platelet activating factor biosynthesis in human neutrophils. 1994; submitted.

5. Miller DK, Gillard JW, Vickers PJ, Sadowski S, Leveille C, Mancini JA, Charleson P, Dixon RAF, Ford-Hutchinson AW, Fortin R, Gautier JY, Rodkey J, Rosen R, Rouzer C, Sigal IS, Strader CD, Evans JF. Identification and isolation of a membrane protein necessary for leukotriene production. Nature 1990; 343:278-281.

6. McDonald PP, McColl SR, Naccache PH, Borgeat P. Studies on the activation of human neutrophil 5-lipoxygenase induced by natural agonists and Ca^{2+} ionophore A23187. Biochem J 1991; 280:379-385.

7. Lazer ES, Wong HC, Wegner CD, Graham AG, Farina PF. Effect of structure on potency and selectivity in 2,6-disubstituted 4-(2-arylethenyl)phenol lipoxygenase inhibitors. J Med Chem 1990; 33:1892-1898.

8. Chilton FH, Connell TR. 1-Ether-linked phosphoglycerides. Major sources of arachidonate in the human neutrophil. J Biol Chem 1988; 263:5260-5265.

9. Chilton FH, Ellis JM, Olson SC, Wykle RL. 1-O-Alkyl-2-arachidonoyl-sn-glycero-3-phosphocholine. A common source of platelet-activating factor and arachidonate in human polymorphonuclear leukocytes. J Biol Chem 1984; 259:12014-12019.

10. Alonso F, Henson PM, Leslie CC. A cytosolic phospholipase in human neutrophils that hydrolyzes arachidonoyl-containing phosphatidylcholine. Biochim Biophys Acta 1986; 878:273-280.

11. Ramesha CS, Ives DL. Detection of arachidonoyl-selective phospholipase A$_2$ in human neutrophil cytosol. Biochim Biophys Acta 1993; 1168:37-44.

12. Wright GW, Ooi CE, Weiss J, Elsbach P. Purification of a cellular (granulocyte) and an extracellular (serum) phospholipase A$_2$ that participate in the destruction of Esherichia coli in a rabbit inflammatory exudate. J BiolChem 1990; 265:6675-6681.

13. Snyder F, Lee TC, Blank ML. The role of transacylases in the metabolism of arachidonate and platelet activating factor. Prog Lipid Res. 1992; 31:65-86.

14. Venable ME, Nieto ML, Schmitt JD, Wykle RL. Evidence that hydrolysis of ethanolamine plasmalogens triggers synthesis of platelet-activating factor via a transacylation reaction. J Biol Chem 1991; 266:18691-18698.

15. Kramer RM, Patton GM, Pritzker CR, Deykin D. Metabolism of platelet-activating factor in human platelets. Transacylase-mediated synthesis of 1-O-alkyl-2-arachidonoyl-sn-glycero-3-phosphocholine. J Biol Chem 1984; 259:13316-13320.

16. Channon JY and Leslie CC. A calcium-dependent mechanism for associating a soluble arachidonoyl-hydrolyzing phospholipase A$_2$ with membrane in the macrophage cell line RAW 264.7. J Biol Chem 1990; 265:5409-5413.

17. Clark JD, Lin L-L, Kriz RW, Ramesha CS, Sultzman LA, Lin AY, Milona N and Knopf JL. A novel arachidonic acid-selective cytosolic PLA$_2$ contains a Ca^{2+}-dependent translocation domain with homology to PKC and GAP. Cell 1991; 65:1043-1051.

18. Lin L-L, Wartmann M, Lin AY, Knopf JL, Seth A, Davis RJ. cPLA$_2$ is phosphorylated and activated by MAP kinase. Cell 1993; 72:269-278.

19. Nemenoff RA, Winitz S, Qian N-X, Van Putten V, Johnson GL, Heasley LE. Phosphorylation and activation of a high molecular weight form of phospholipase A$_2$ by p42 microtubule-associated protein 2 kinase and protein kinase C. J Biol Chem 1993; 268:1960-1964.

20. Vadas P, Pruzanski W. Role of secretory PLA$_2$ in the pathobiology of disease. Lab Invest 1986; 55:391-404.

21. Elsbach P, Weiss J. Phagocytosis of bacteria and phospholipid degradation. Biochim Biophys Acta 1988; 947:29-52.

22. Fonteh AN, Bass DA, Marshall LA, Seeds M, Samet JM, Chilton FH. Evidence that secretory phospholipase A$_2$ plays a role in arachidonate release and eicosanoid biosynthesis by mast cells. J Immunol 1994; 152:5438-5446.

23. Jones DS, Seeds MC, Chilton FH, Bass DA. Priming of two phospholipases A$_2$ of human neutrophils by tumor necrosis factor. 1994; submitted.

AAS 46
Novel Molecular Approaches
to Anti-Inflammatory Theory
© 1995 Birkhäuser Verlag Basel

THE USE OF ANTI-PECAM REAGENTS IN THE CONTROL OF INFLAMMATION

William A. Muller

The Rockefeller University, 1230 York Avenue, New York, NY 10021

SUMMARY: Platelet/endothelial cell adhesion molecule 1 (PECAM-1/CD31) is expressed on the surfaces of neutrophils and monocytes and concentrated at the junctional surfaces of vascular endothelial cells. Monoclonal antibodies against PECAM-1 and soluble recombinant PECAM-1 selectively block the passage of these leukocytes across the endothelial monolayer without interfering with earlier adhesion events in the emigration pathway. This block is seen both *in vitro* and in several *in vivo* models of acute inflammation. Since PECAM-1 appears to be crucial for a distinct step in the emigration of leukocytes into a focus of inflammation, PECAM-1 appears to be a new and potentially important target for anti-inflammatory therapy.

INTRODUCTION

At a site of inflammation, circulating leukocytes leave the bloodstream from postcapillary venules (1) and enter the affected tissues. As observed by intravital microscopy more than a century ago (2), leukocytes leave the more rapidly flowing stream of blood and begin to bind weakly and reversibly to the luminal surface of the endothelial cells lining these vessels. As a result, these leukocytes slow down and "roll" along the vascular wall. Shortly thereafter, certain of these white cells will suddenly adhere tightly to the endothelial surface; they cease rolling and spread on the endothelial cell surface. Contemporaneously, they migrate to a nearby intercellular junction and leave the circulation by squeezing between tightly apposed endothelial cells (3).

We are now beginning to understand the molecular mechanisms underlying these events. The emigration phenomenon has been dissected into four distinct steps by use of antibodies and recombinant reagents that block the various stages. Each of these steps -- rolling, activation, tight adhesion, and transmigration -- appears to be mediated predominantly by a distinct family of cell adhesion molecules (CAMs). These are integral membrane proteins, exposed on the cell surface, that mediate adhesion by binding to a ligand or counter-receptor on the apposing cell. The rolling phenomenon involves interaction of members of the selectin family with their sialylated-Lewis^x-decorated ligands on the apposing cell (4,5). Leukocytes bear L-selectin constitutively; endothelial cells can be induced to rapidly degranulate preformed P-selectin following stimulation by thrombin or histamine (6). E-selectin is transcribed *de novo* from mRNA induced by cytokine activation of endothelium; maximal expression takes several hours (7). The rolling process brings leukocytes into contact with the endothelial surface, so that they can be exposed to one or more of a variety of signals that will activate tight adhesion via the leukocyte integrins (5,8,9). Once activated, the leukocyte integrins appear to mediate the tight adhesion and spreading of the white cells to the endothelial surface as well as their locomotion along the surface to the intercellular junctions (5,8,9). Those ligands on endothelial cells that have been identified for the leukocyte integrins are members of yet another CAM family, the immunoglobulin gene superfamily, and include ICAM-1, ICAM-2, and VCAM-1 (10).

The transmigration step has been the least studied, but one molecule that appears to be critical for the emigration of neutrophils and monocytes is platelet/endothelial cell adhesion molecule 1 (PECAM-1/CD31). This is a 130-kd member of the immunoglobulin gene superfamily (11) that is expressed on the surfaces of neutrophils and monocytes as well as on endothelial cells (12), where it is concentrated at the lateral borders of the cells. Monoclonal antibody against PECAM-1 as well as soluble recombinant PECAM-1 have been used to selectively block transendothelial migration of both neutrophils and monocytes in an *in vitro* model (13). Monoclonal (14) and polyclonal (15) antibodies to PECAM-1 have been used to block acute inflammation in rodent models.

RESULTS AND DISCUSSION

PECAM-1 can mediate homophilic (PECAM binding to PECAM) (12,16,17) as well as heterophilic (18) (PECAM binding to sulfated glycosaminoglycans [19]) adhesion, depending on the experimental conditions. My laboratory had developed a quantitative *in vitro* assay of transendothelial migration in which monocytes and neutrophils could be induced to transmigrate human umbilical vein endothelial cell monolayers with kinetics similar to those reported from *in vivo* studies (20). Since PECAM is expressed on the surfaces of monocytes and neutrophils as well as at the endothelial cell junctions, we tested the hypothesis that PECAM was important in the process of transendothelial migration.

When monocytes are preincubated with anti-PECAM monoclonal antibody (mAb) hec7, they bind tightly to the apical endothelial cell surface and migrate to the junction, but they do not transmigrate. They appear to be blocked at a step distal to the tight integrin-mediated adhesion, since they resist inverted centrifugation in EGTA (20). Our quantitative assay demonstrated that the anti-PECAM treatment did not block adhesion per se; the same number of leukocytes remained tightly associated with the monolayer as in controls. What was different was the position of these bound leukocytes. Light- and scanning electron microscopy demonstrated that the leukocytes were apparently arrested at the intercellular junction; many demonstrated a small pseudopod inserted into the junction (13).

The existing assay was modified to quantitate this phenomenon. We took advantage of the fact that monocytes and neutrophils bear on their surfaces receptors for the Fc domain of IgG. Furthermore, this receptor functions at 4°C. At the end of the assay, the experimental samples were washed with cold buffer and chilled. Immunoglobulin-coated sheep erythrocytes were allowed to settle onto the monolayers for thirty minutes, then the monolayer surfaces were washed. Leukocytes that were arrested on the apical surface of the endothelial monolayer bound these red cells and could be seen as "rosettes" on the monolayer surface. Those leukocytes that had transmigrated the monolayer could not come into contact with the red cells and thus were not rosetted. Non-rosetted leukocytes were counted as transmigrated. As measured by this assay, transendothelial migration was blocked by 70%-90% by concentrations of mAb hec7 as low as 10 μg/ml; Fab fragments worked well, too, so this effect was not mediated via binding of hec7 to the monocyte Fc receptor. Anti-PECAM reagents blocked migration of monocytes and neutrophils (PMN) across cytokine-activated endothelial cell monolayers just as well as they blocked transmigration across resting endothelial cell monolayers. (See Fig. 1 and reference

[13]). The effect lasted for at least 6 hours in the continued presence of antibody and was reversible once antibody was withdrawn. Control experiments demonstrated that the monoclonal antibody did not affect the ability of leukocytes to move or to respond to a chemotactic gradient and that the effect was selective for antibodies that bound to PECAM (13).

Transmigration could also be blocked by anti-PECAM reagents in the intercellular junction. The block in transmigration was equivalent whether antibody was added to the leukocyte or the endothelial cell side, and blocking PECAM on both cells did not augment the effect, an observation suggesting that PECAM on the leukocyte was binding to PECAM on the endothelial cell. Moreover, soluble recombinant PECAM had the same effect as monoclonal antibody; thus it appeared that a competitive inhibition was taking place (13).

Figure 1. **Anti-PECAM reagents significantly block transendothelial migration of both monocytes and neutrophils (PMN).** Leukocytes were exposed to the indicated anti-PECAM reagents and then incubated with resting (Monocytes) or TNFα-activated (PMN) human umbilical vein endothelial cell monolayers for 1 h. at 37°C. Monolayers were washed, and the percentage of transmigrated cells assayed as described in the text and in (13). The mean ± S.E.M. of six replicate samples is shown. Sera were diluted 1:100, monoclonal antibodies were used at 20 μg/ml, soluble recombinant PECAM-1 was used at 10 μg/ml. Note that although monoclonal antibody hec7 significantly blocked transmigration of PMN, it did not do so as well as it blocked transmigration of monocytes at that concentration. This may be due to subtle differences between the PECAM of monocytes and that of PMN at the epitope recognized by hec7, since other anti-PECAM reagents were just as effective at blocking PMN as monocytes.

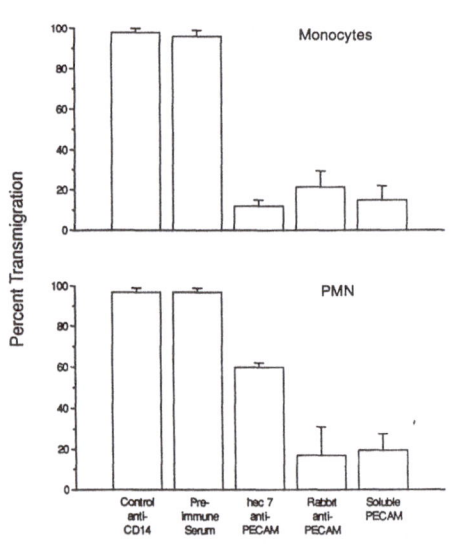

To test the role of this molecule in an *in vivo* model of inflammation, we took advantage of the fact that the murine homologue of PECAM-1 had recently been described. Cloning of the murine homologue of PECAM-1 (21) revealed a predicted amino acid sequence with 79%

homology to human PECAM-1. L cells transfected with murine PECAM-1 cDNA aggregated in a PECAM-1-dependent manner, similar to human PECAM-1 transfectants. A monoclonal antibody (mAb 2H8) raised in hamsters that recognizes the murine form of PECAM-1 (22) blocked this aggregation (21). Furthermore, immunohistochemical studies using this antibody demonstrated that murine PECAM-1 had a tissue distribution similar to that of human PECAM-1 (22).

We used a murine model of acute peritonitis to test whether intravenously (i.v.) administered mAb 2H8 would block acute inflammation (14). In this well-established model of acute peritonitis, intraperitoneal thioglycollate induces an influx of neutrophils into the peritoneal cavity within the first two hours (23,24). The degree of inflammation was measured by peritoneal lavage 20 hours after thioglycollate injection. Two representative experiments are shown in Fig. 2 and 3. In all cases, the effect of mAb 2H8 was comparable to that of the positive control mAb 5C6 (anti-CD11b), which had previously been shown to block acute inflammation in this model (25).

Figures 2 and 3 separately report data on the PMN exudate and the total exudate cells (including lymphocytes and macrophages). The decrease in PMN accumulation produced by the mAb 2H8 is particularly dramatic, since these cells are not normally resident in the peritoneal cavity. Significant suppression of inflammation by mAb 2H8 was achieved at doses as low as 50 μg/mouse (the lowest dose tested), and suppression was maintained for at least 48 hours (not shown). This time course is similar to that described with anti-CD11b blockade (25) and significantly longer than the four-hour suppression seen in this model by blockade of L-selectin (24) or knockout of P-selectin (26). Direct enumeration of peripheral blood leukocytes demonstrated that the block in leukocyte emigration observed in the mAb 2H8-treated mice was not due to a decrease in circulating cells. Control experiments also ruled out that this was mediated by the Fc domain of the antibody, but rather, it was a direct effect of the antigen-recognizing domain. (Fig. 3 and reference [14].)

Histologic examination of the mesenteries of the mice receiving mAb 2H8 showed increased numbers of intravascular PMN in the venules compared with controls (14). These PMN appeared to be in contact with the endothelial surface but were inhibited from migrating across the vascular wall, analogous to the block of transmigration observed *in vitro* (13). Only rare

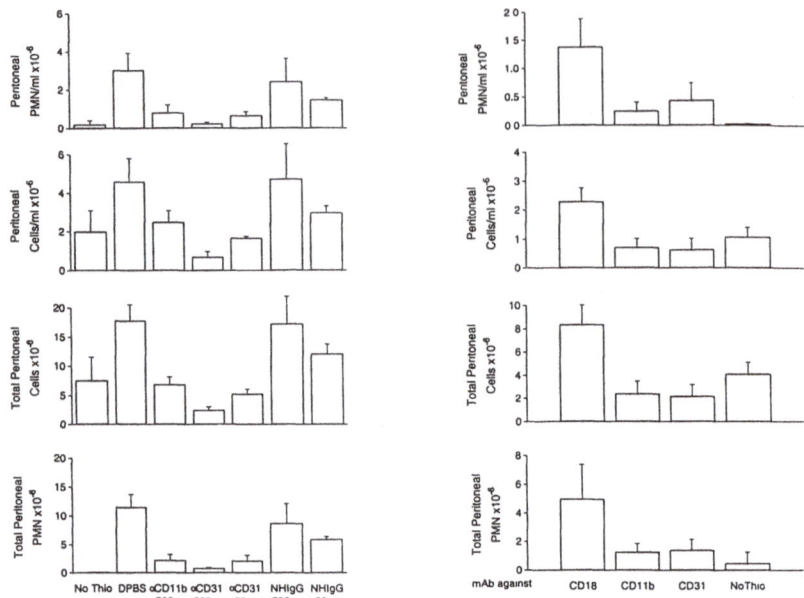

Figure 2. (left) **Anti-PECAM mAb blocks leukocyte emigration.** Mice were injected via lateral tail vein with the indicated dose of mAb, normal hamster IgG (NHIgG), or Dulbecco's PBS (DPBS) in 0.1 ml of DPBS. Four hours later all mice except the "No Thio" controls were injected intraperitoneally with 1 ml of thioglycollate, as described in Methods. Peritoneal cells were harvested 20 hours after the thioglycollate challenge. 2H8 = hamster anti-mouse PECAM (CD31); 5C6 = rat anti-mouse Mac-1 (CD11b). The graphs depict the mean and standard deviation for each group. (Three mice per group.)

Figure 3. (right) **Anti-PECAM mAb blocks leukocyte emigration.** The experiment was performed similarly to the one in Fig. 1, but each experimental group contained **five mice**, and each mouse received 250 μg of mAb intravenously prior to thioglycollate stimulation. Control mice (No thio) received DPBS i.v. Monoclonal antibody 2E6 is a hamster IgG (like 2H8) and binds to leukocyte CD18, but does not block function, and thereby demonstrates that mere binding of a hamster IgG to mouse leukocytes does not interfere with emigration of PMN. Similar numbers were obtained with other negative controls.

Peripheral blood leukocyte counts and differentials for this experiment were:

mAb	WBC (cells/μl)	Differential (%PMN/ %Lymph/ %Mono) (mean)
2E6	9,640 ± 1,270	28/65/7
5C6	11,580 ± 3,460	20/74/6
2H8	16,060 ± 2,750	59/33/8
-	12,260 ± 1,970	11/77/12

Note that unlike humans, a normal mouse WBC differential (-) is dominated by lymphocytes. Mice receiving thioglycollate displayed a "shift to the left". Immature forms are recorded here as PMN, for simplicity.

leukocytes contacted the endothelium in control (no thioglycollate) animals or in animals receiving mAb other than 2H8. Thirty four percent of all leukocytes seen in the venules of the mAb 2H8-treated mice were in apparent contact with the endothelium. Immunohistologic examination of these mice demonstrated that mAb 2H8 was still specifically retained by endothelium in vascular structures, including mesenteric venules. We presume that the arrest on the endothelial surface is transient, since the vast majority of the luminal surface of the affected venules was free of leukocytes, and the granulocytosis was persistent. The block of emigration effected by mAb 2H8 appears to involve a different mechanism than the block by anti-CD11b. In the latter case, no granulocytosis was seen in the peripheral blood, and, although PMN emigration into the peritoneum was inhibited, leukocytes were not seen in apparent contact with the venular wall. This is consistent with the proposed role of β_2 integrins in mediating the tight adhesion to the endothelial surface (8,27,28).

Significant numbers of mononuclear cells as well as PMN are recruited to the peritoneal cavity by 20 hours after thioglycollate injection. Monoclonal antibody 2H8 also blocked the influx of mononuclear cells into the peritoneum (Figs. 2 and 3 and reference [14]). Furthermore, in some experiments, the thioglycollate-elicited peritoneal exudate cell count was lower for mAb 2H8-treated mice than for control mice not stimulated with thioglycollate. This suggested that PECAM-1 may be required for the constitutive trafficking of mononuclear cells through the peritoneum as well as for the thioglycollate-elicited emigration.

In our *in vitro* model, leukocyte transmigration could be blocked equally well by treatment of either the leukocytes or the endothelial cells with anti-PECAM-1 reagents. We do not know whether mAb 2H8 is blocking leukocyte transmigration by blocking PECAM-1 on leukocytes, endothelium, or both.

Since mAb 2H8 blocked emigration of mononuclear cells as well as PMN *in vivo*, PECAM-1 may mediate a function common to all leukocyte types that is necessary for the process of transendothelial migration. In addition, the inhibition was effected at a low dose of mAb 2H8 (50 μg/mouse in certain experiments) and lasted up to 48 hours following a single injection. Direct immunohistologic observation of these tissues demonstrated that mAb 2H8 was selectively bound to vascular endothelium 24 hours after injection (14). These observations suggest that PECAM-1 may be a suitable molecule to target in anti-inflammatory therapy.

The *in vivo* efficacy of anti-PECAM reagents has been confirmed in other species and models. Using intravenously administered polyclonal rabbit antibody against human PECAM-1 that recognized rat PECAM, Vaporciyan et al. (15) were able to block neutrophil entry into the peritoneal cavity of rats following i.p. injection of glycogen. The same antibody significantly reduced neutrophil influx into the alveoli 4 hours after intratracheal installation of IgG immune complexes. Furthermore, it blocked TNFα-induced emigration of murine leukocytes into grafts of human foreskin transplanted onto SCID mice. Although these studies used polyclonal antibody and were conducted over a comparatively short time (4 hours), they support the premise that anti-PECAM reagents may be used successfully in a variety of inflammatory settings.

Since PECAM-1 is expressed on leukocytes and endothelial junctions, and since anti-PECAM reagents directed against either side are equally effective at blocking transmigration *in vitro* (13), it is possible that systemic anti-PECAM therapy could be used in the case of a generalized inflammatory state (e.g. systemic lupus erythematosus), while selective infusion of anti-PECAM therapy into a restricted vascular bed could be useful in preventing graft rejection or in treating monarticular arthritis. There are subtle, but detectable differences in PECAM-1 expressed on platelets, endothelial cells, neutrophils, and monocytes (See Fig. 1 and reference [11].) We may be able to capitalize on these differences to make anti-PECAM reagents that would interfere selectively with PECAM function on these cell types.

Emigration of leukocytes into a site of inflammation involves several sequential adhesion events, as outlined in the Introduction. Currently, anti-inflammatory therapies are targeting the individual molecules -- especially the selectins and integrins -- involved in these steps. The major function of PECAM-1 appears to be distal to the tight adhesion step and independent of the previous stages and adhesion molecules (13). Therefore, although anti-PECAM reagents appear to be useful anti-inflammatory reagents *in vivo* on their own, they may also be used in conjunction with reagents directed against the other stages to improve the overall anti-inflammatory activity.

ACKNOWLEDGEMENTS

Supported by Grants HL46849 and AI24775 from the National Institutes of Health, United States Public Health Service.

REFERENCES

1. Marchesi VT. The site of leukocyte emigration during inflammation. J Exp Physiol 1961; 46:115-118.

2. Cohnheim J. Lectures on general pathology: a handbook for practitioners and students. London, England: The New Sydenham Society, 1889:

3. Marchesi VT,Florey HW. Electron micrographic observations on the emigration of leukocytes. J Exp Physiol 1960; 45:343-347.

4. Lasky LA. Selectins: Interpreters of cell-specific carbohydrate information during inflammation. Science 1992; 258:964-969.

5. Springer TA. Traffic signals for lymphocyte recirculation and leukocyte emigration: The multistep paradigm. Cell 1994; 76:301-314.

6. Bevilacqua MP,Nelson RM. Selectins. J Clin Invest 1993; 91:379-387.

7. Bevilacqua MP, Stengalin S, Gimbrone MA,Seed B. Endothelial leukocyte adhesion molecule 1: an inducible receptor for neutrophils related to complement regulatory proteins and lectins. Science 1990; 243:1160-1165.

8. von Andrian UH, Chambers JD, McEvoy LM, Bargatze RF, Arfors K-E,Butcher EC. Two-step model of leukocyte-endothelial cell interaction in inflammation: Distinct roles for LECAM-1 and the leukocyte beta 2 integrins in vivo. Proc Natl Acad Sci USA 1991; 88:7538-7542.

9. Lorant DE, Patel KP,McIntyre TM. Coexpression of GMP-140 and PAF by endothelium stimulated by histamine or thrombin: a juxtacrine system for adhesion and activation of neutrophils. J Cell Biol 1991; 115:223-234.

10. Springer TA. Adhesion receptors of the immune system. Nature (London) 1990; 346:425-434.

11. Newman PJ, Berndt MC, Gorski J, et al. PECAM-1 (CD31) cloning and relation to adhesion molecules of the immunoglobulin gene superfamily. Science 1990; 247:1219-1222.

12. Muller WA, Ratti CM, McDonnell SL,Cohn ZA. A human endothelial cell-restricted,

externally disposed plasmalemmal protein enriched in intercellular junctions. J Exp Med 1989; 170:399-414.

13. Muller WA, Weigl SA, Deng X,Phillips DM. PECAM-1 is required for transendothelial migration of leukocytes. J Exp Med 1993; 178:449-460.

14. Bogen S, Pak J, Garifallou M, Deng X,Muller WA. Monoclonal antibody to murine PECAM-1 (CD31) blocks acute inflammation in vivo. J Exp Med 1994; 179:1059-1064.

15. Vaporciyan AA, Delisser HM, Yan H-C, et al. Involvement of platelet-endothelial cell adhesion molecule-1 in neutrophil recruitment in vivo. Science 1993; 262:1580-1582.

16. Albelda SM, Muller WA, Buck CA,Newman PJ. Molecular and cellular properties of PECAM-1 (endoCAM/CD31): A novel vascular cell-cell adhesion molecule. J Cell Biol 1991; 114:1059-1068.

17. Muller WA. PECAM-1: an adhesion molecule at the junctions of endothelial cells. In: van Furth R, Cohn ZA, Gordon S, eds. Mononuclear Phagocytes. The Proceedings of the Fifth Leiden Meeting on Mononuclear Phagocytes. London: Blackwell Publishers, 1992:148.

18. Muller WA, Berman ME, Newman PJ, Delisser HM,Albelda SM. A heterophilic adhesion mechanism for Platelet/Endothelial Cell Adhesion Molecule-1 (CD31). J Exp Med 1992; 175:1401-1404.

19. Delisser HM, Yan HC, Newman PJ, Muller WA, Buck CA,Albelda SM. Platelet/endothelial cell adhesion molecule-1 (CD31)-mediated cellular aggregation involves cell surface glycosaminoglycans. J Biol Chem 1993; 268:16037-16046.

20. Muller WA,Weigl S. Monocyte-selective transendothelial migration: Dissection of the binding and transmigration phases by an in vitro assay. J Exp Med 1992; 176:819-828.

21. Xie Y,Muller WA. Molecular cloning and adhesive properties of murine platelet/endothelial cell adhesion molecule-1. Proc Natl Acad Sci USA 1993; 90:5569-5573.

22. Bogen SA, Baldwin HS, Watkins SC, Albelda SM,Abbas AK. Association of murine CD31 with transmigrating lymphocytes following antigenic stimulation. Am J Pathol 1992; 141:843-854.

23. Lewinsohn DM, Bargatze RF,Butcher EC. Leukocyte-endothelial cell recognition: Evidence of a common molecular mechanism shared by neutrophils, lymphocytes, and other leukocytes. J Immunol 1987; 138:4313-4321.

24. Watson SR, Fennie C,Lasky LA. Neutrophil influx into an inflammatory site inhibited by a soluble homing receptor-IgG chimaera. Nature (London) 1991; 349:164-167.

25. Rosen H,Gordon S. Monoclonal antibody to the murine type 3 complement receptor inhibits adhesion of myelomonocytic cells in vitro and inflammatory cell recruitment in vivo. J Exp Med 1987; 166:1685-1701.

26. Mayadas TN, Johnson RC, Rayburn H, Hynes RO,Wagner DD. Leukocyte rolling and extravasation are severely compromised in P selectin-deficient mice. Cell 1993; 74:541-554.

27. Lawrence MB,Springer TA. Leukocytes roll on a selectin at physiologic flow rates: Distinction from and prerequisite for adhesion through integrins. Cell 1991; 65:859-873.

28. Lo SK, Lee S, Ramos RA, et al. Endothelial-leukocyte adhesion molecule 1 stimulates the adhesive activity of leukocyte integrin CD3 (CD11B/CD18, Mac-1, alpha m beta 2) on human neutrophils. J Exp Med 1991; 173:1493-1500.

AAS 46
Novel Molecular Approaches
to Anti-Inflammatory Theory
© 1995 Birkhäuser Verlag Basel

SELECTIVE INHIBITORS OF COX-2

G.P. O'Neill, B.P. Kennedy, J.A. Mancini, S. Kargman, M. Ouellet, J. Yergey, J.-P. Falgueyret, W.A. Cromlish, P. Payette, C.-C. Chan, S.A. Culp, C. Vincent, C. Boily, M. Abramovitz, J.F. Evans, A.W. Ford-Hutchinson, P.J. Vickers, and M.D. Percival

Merck Frosst Centre for Therapeutic Research, P.O. Box 1005, Pointe-Claire-Dorval, Québec, H9R 4P8, Canada

SUMMARY: The main target of non-steroidal anti-inflammatory drugs (NSAIDs) is prostaglandin G/H synthase (PGHS), also known as cyclooxygenase (COX), which exists as two isoforms. In order to evaluate the contributions of PGHS isoforms to physiological and pathological conditions and their sensitivity to inhibition by non-steroidal anti-inflammatory drugs, we have established high level expression systems of recombinant human PGHS isoforms. The inducible form of PGHS, termed PGHS-2, has been purified and characterized with respect to substrate specificity, product formation, enzymatic activity, glycosylation, heme content, quaternary structure, and modification by aspirin. Pharmacolgical profiles of the recombinant PGHS isoforms indicate that conventional NSAIDs show little selectivity for either enzyme, however, the recently described NSAID, NS-398, exhibits a high degree of specificity for PGHS-2 through a time dependent mechanism.

INTRODUCTION

Prostaglandin G/H synthase (PGHS), also known as cyclooxygenase (COX), catalyzes the conversion of arachidonic acid to prostaglandin H_2, the common precursor for the formation of prostanoids. Non-steroidal anti-inflammatory drugs (NSAIDs) exert their anti-inflammatory activity by inhibition of PGHS, which occurs in two isoforms, termed PGHS-1 and PGHS-2 (1-5). These isoforms differ in several respects, including their regulation of expression (1-5), primary amino acid sequence (4,5), enzymatic activity following acetylation by aspirin (6,7), and differential inhibition by NSAIDs (7-10). PGHS-1 is constitutively expressed in a wide variety of cell types, in contrast to PGHS-2 which has been shown to be rapidly induced in many cell types by a number of agents including cytokines, phorbol esters, endotoxin, serum factors, and mitogens (1-3,5). This rapid and high level of induction of PGHS-2 expression by pro-inflammatory agents has led to the hypothesis that PGHS-2 plays a central role in mediating the rapid onset of the inflammatory response, whereas the constitutive expression of PGHS-1 provides prostanoids for the normal physiology of cells and tissues (8-11).

The discovery of a second form of PGHS expressed in inflammation has generated much interest in the possibility of whether PGHS-2 isozyme selective NSAIDs may be developed that prove beneficial in treating a variety of inflammatory conditions with reduced potential for gastrointestinal and renal toxicity (7,8,10,11). Indeed, under certain assay conditions classical NSAIDs have been reported to exhibit differential inhibition of PGHS isoforms (8,10), however, our studies show that classical NSAIDs do not show a high level of selectivity for either PGHS isoform (7). Recently, a new anti-inflammatory and analgesic agent, NS-398, has been reported to selectively inhibit PGHS-2 activity in vitro (12,13). In rat models of inflammation NS-398 has been shown to be a potent antiinflammatory agent but produces much fewer gastrointestinal lesions than the conventional NSAID indomethacin (9,12,13).

To study the characteristics of PGHS isoforms and the NSAID inhibition of their enzyme activities, it is critical to establish individual sources of the isozymes. The present work describes the high-level expression of recombinant human PGHS isoforms in both mammalian cells using vaccinia virus and insect cells using baculovirus expression vectors (7,14). The products of the recombinant PGHS-1 and -2 catalyzed reactions have been characterized. In addition, data on the inhibition of product formation by the PGHS isoforms by standard NSAIDs has been obtained (7,14). The recombinant human PGHS-2 proteins produced using either the vaccinia virus or baculovirus expression systems have been purified to homogeneity and characterized with respect to substrate specificity, product formation, enzymatic activity, glycosylation, heme content, and quaternary structure (15). We have also investigated the kinetic mechanism of inhibition of purified human PGHS-1 and PGHS-2 by NS-398 and four classical NSAIDs (16). Furthermore, we have examined the aspirin-stimulated formation of 15-R-hydroxyteicosatetraenoic acid (15-R-HETE) by hPGHS-2 (7,17) and the effects of NSAIDs on aspirin-stimulated 15-R-HETE formation by hPGHS-2 (18). In an attempt to mimic the effects of aspirin acetylation on hPGHS-2 activity we mutated the putative aspirin acetylation site of the enzyme by site-directed mutagenesis and assessed the effects of amino acid substitutions on enzyme activity (17).

RECOMBINANT OVEREXPRESSION OF hPGHS-1 BY VACCINIA VIRUS

Studies on the recombinant expression of PGHS-1 and -2 in transiently transfected mammalian cell lines and expression of sheep PGHS-1 using a baculovirus system suggest that post-translational glycosylation of PGHS isoforms is essential in obtaining high specific activity enzyme (19,20). Since the vaccinia virus (VV) system is widely used to produce large amounts of proteins that undergo correct post-translational modifications in a broad range of mammalian host cells, we chose this system for the high level expression of the PGHS isoforms (7). The hPGHS-1 cDNA, which encodes a 599 amino acid protein containing a putative 23-amino acid leader peptide and four potential glycosylation sites (4), was transferred by genetic recombination into vaccinia virus to yield the recombinant VV:hPGHS-1 (7). In the T7 RNA polymerase/vaccinia virus expression system utilized here, the hPGHS-1 cDNA sequence is located downstream from a T7 RNA polymerase promoter sequence. Expression of the recombinant hPGHS-1 requires coinfection with two recombinant viruses (21). One recombinant virus, VV:TF7-3, contains the bacteriophage T7 RNA polymerase gene under the control of the VV P7.5 promoter. The second recombinant virus contains the hPGHS sequences flanked by the T7 promoter (7).

COS-7 cells, which almost completely lack detectable endogenous PGHS activity, were coinfected with VV:TF7-3 and VV:hPGHS-1 and then analyzed for the expression of hPGHS-1 by immunoblot analysis and enzyme activity assays (7). Immunoblot analysis demonstrated the expected 72-kDa protein corresponding to glycosylated PGHS-1 only in membrane preparations from COS-7 cells coinfected with VV:TF7-3 and VV:hPGHS-1. The PGHS activity in these infected cells was >1500-fold higher than in mock-infected COS-7 cells and was approximately 75-fold higher than in microsomes prepared from the human histiocytic lymphoma cell line U-937, which has been shown to express hPGHS-1. The GC/MS profiles of prostanoids produced by cells coinfected with VV:TF7-3 and VV:hPGHS-1 demonstrated that prostaglandin E_2 (PGE_2) and prostaglandin D_2 (PGD_2) were the most abundant PGs produced. Treatment of cells coinfected with VV:TF7-3 and VV:hPGHS-1 with tunicamycin, an inhibitor of N-linked glycosylation, yielded microsomal fractions lacking PGHS activity, although these microsomes contained an equivalent amount of immunoreactive PGHS-1 protein with a 4-kDa reduction in apparent molecular mass. These results suggested that expression of hPGHS-1 requires glycosylation of the enzyme to obtain full enzymatic activity.

RECOMBINANT OVEREXPRESSION OF hPGHS-2 BY VACCINIA VIRUS

The hPGHS-2 cDNA encodes a 604 amino acid polypeptide that is 61% identical to hPGHS-1, and contains an 18-amino acid signal peptide, five potential glycosylation sites, and numerous residues proposed to be at the active site or involved in heme coordination that are conserved in hPGHS-1 (5). It should be noted that the hPGHS-2 cDNA that we have used in these studies differs from the published hPGHS-2 sequence at amino acid residue 165 (5,14). The previously reported hPGHS-2 cDNA has a glycine at position 165 as compared to our sequence which has a glutamic acid at this position. Expression studies of hPGHS-2$_{glu165}$ and hPGHS-2$_{gly165}$ in transiently transfected COS-7 cells demonstrated that the hPGHS-2$_{glu165}$ encodes for a protein with approximately 2-3-fold higher activity than hPGHS-2$_{gly165}$ (14). The hPGHS-2 cDNA is distinguished from the hPGHS-1 cDNA by the presence in its 3' untranslated region (UTR) of over 12 copies of the Shaw-Kamen motif found in many immediate-early genes (5). In comparison the 3' UTR's of PGHS-1 from several species show an unusually high level of sequence homology and the potential to form stable secondary structures (4). Since the Shaw-Kamen motifs found in the hPGHS-2 3' UTR have been shown to confer enhanced mRNA degradation, the hPGHS-2 open reading frame (ORF) without its 3' UTR was used to construct the expression vector VV:hPGHS-2-orf (7). However, when COS-7 cells coinfected with VV:TF7-3 and VV:hPGHS-2-orf were examined only low levels of PGHS-2 mRNA and protein expression were detected. Because of the previous suggestion that the ORF of hPGHS-2 mRNA may contain instability elements (5), we attempted to stabilize the hPGHS-2 mRNA by appending the 3' UTR of hPGHS-1 to the ORF of hPGHS-2, in a VV designated VV:hPGHS-2-3'fl (7). Immunoblot analysis of membranes prepared from COS-7 cells coinfected with VV:TF7-3 and VV:hPGHS-2-3'fl demonstrated the overexpression of two proteins of 72 and 74-kDa which co-migrated with purified sheep PGHS-2. The amount of recombinant hPGHS-2 in these microsomes was approximately 4% of the total microsomal protein. As with recombinant hPGHS-1, GC/MS data confirmed PGE_2 as the predominant prostanoid formed in cells infected with VV:hPGHS-2-3'fl. In addition, tunicamycin blocked post-translational glycosylation of the recombinant protein and yielded an enzymatically inactive hPGHS-2 of 69 kDa.

RECOMBINANT OVEREXPRESSION OF hPGHS-2 IN INSECT CELLS BY BACULOVIRUS

Although the VV expression system provides very high yields of glycosylated, active hPGHS-1 and -2, we have also investigated the production of hPGHS-2 by the baculovirus (BV) expression system because of the ease of scale up using this system and the potential safety issues associated with the VV system (14). Thus, the hPGHS-2 ORF was recombined into baculovirus using the transfer vector pVL 941 and expression studies of hPGHS-2 in Sf9 insect cells were carried out (14). Immunoblot and enzyme activity analysis of Sf9 insect cells infected with the hPGHS-2 recombinant baculovirus vector, designated BV:hPGHS-2-orf, revealed a tremendous increase in the amount of PGHS-2 immunoreactive protein and PGHS activity in the low speed nuclear pellet and the high speed microsomal pellet. PGHS activity in Sf9 insect cell cultures infected with BV:hPGHS-2-orf peaked at about 48 hours post-infection and remained elevated up to 79 hours. The specific activity of the hPGHS-2 in microsomes prepared from the infected Sf9 cells was 0.51 μmole O_2/min/mg protein and was nearly identical to the specific activity of purified sheep PGHS-2 (0.55 μmole O_2/min/mg protein) and that of microsomal hPGHS-2 prepared from COS-7 cells infected with VV:hPGHS-2-3'fl as discussed above (7,14). It is interesting to note that the hPGHS-2 ORF was sufficient to direct overexpression of this protein in the BV system in contrast to the VV system which required a 3' UTR for high level expression.

Immunoblot analysis showed that the BV-derived hPGHS-2 was a heterogenous mixture of differentially glycosylated proteins of molecular mass between 66-72 kDa, in contrast to the VV-derived hPGHS-2 which was expressed as a doublet at 72-74 kDa (7,14). Computer analysis of the PGHS-2 peptide predicts five potential glycosylation sites, while experimental evidence suggests that three or four of these sites are naturally glycosylated (5,20). In many natural systems PGHS-2 appears as a doublet of 72-74 kDa on sodium dodecyl sulfate-polyacrylamide gels (SDS-PAG). This doublet has been shown to be due to the differential *N*-glycosylation of the asn580 site in 50% of the molecules (20). Site-specific mutagenesis studies have shown that glycosylation of the asn580 site in PGHS-2 is not required to attain enzyme activity, but glycosylation at two or three other sites is absolutely required for the proper processing and folding of the enzyme (20). Our studies indicate that although the glycosylation of hPGHS-2 in the baculovirus system is more heterogeneous than in the vaccinia virus system, the differentially glycosylated BV-derived hPGHS-2 has a specific activity equivalent to the VV-derived hPGHS-2 and purified sheep PGHS-2 (14). The high yields of active hPGHS-2 using the baculovirus system and the ease with which the baculovirus system can be adapted to large bioreactors opens the possibility of producing sufficient quantities of hPGHS-2 for the determination of its X-ray crystal structure.

PURIFICATION AND CHARACTERIZATION OF RECOMBINANT hPGHS-2

In order to investigate interactions between PGHS isoforms and their targets, and to structurally characterize hPGHS-2, we have purified the baculovirus-expressed and vaccinia virus-expressed recombinant hPGHS-2 (15). In both the Sf9 insect cells and COS-7 cells expressing hPGHS-2, the enzyme activity was localized to the membrane fraction obtained following cell lysis and high speed centrifugation at 100,000 X g. Greater than 85% of the PGHS activity could be extracted from the microsomal protein by treatment with 45 mM ß-octylglucoside. Interestingly, the octylglucoside treatment did not solubilize the nonglycosylated hPGHS-2 contained in the baculovirus-derived pellet. The solubilized enzymes were purified to homogeneity using a DEAE-ion exchange HPLC column. The yields of purified hPGHS-2 from the BV and VV expression systems were similar; 1.0 mg of hPGHS-2 was purified from 1.6×10^8 VV:hPGHS-2-3'fl infected COS-7 cells and 7 mg of pure hPGHS-2 was obtained from 1.1×10^9 Sf9 insect cells. As was observed in the crude microsomal preparations of BV- and VV-derived hPGHS-2, the purified enzymes were a heterogeneous mixture of glycoforms. This was confirmed by the observation that treatment of of the SDS-denatured proteins with endoglycosidase H or glycopeptidase F resulted in increased electrophoretic mobility and a shift to a single species (15).

The specific cyclooxygenase activities of the purified hPGHS-2, determined by measuring oxygen uptake using an oxygen electrode and arachidonic acid as substrate, were 43 µmol O_2/min/mg for both BV-derived and VV-derived hPGHS-2. The K_m value of arachidonic acid for both BV- and VV-derived hPGHS-2 was 0.9 µM. Six other C-18 and C-20 fatty acids containing at least one 1,4-cis,cis-pentadiene moiety were identified as substrates for hPGHS-2. These fatty acids included linoleic acid, linolenic acid, γ-linolenic acid, Mead acid, dihomo-γ-linolenic acid, 11,14-eicosadienoic acid, and 5,8,11,14,17-eicosapentaenoic acid. Linoleic and γ-linolenic acid were determined by mass spectrometry as being hydroxylated primarily at the C-9 and C-13 positions, whereas linolenic was hydroxylated primarily at the C-12 and C-16 positions (15).

A spectrophotometric titration of purified BV-derived hPGHS-2 with hematin indicated that the purified enzyme binds heme such that maximal activity is observed at a stoichiometry of 1 heme per enzyme subunit. The apparent molecular mass of hPGHS-2, determined by gel filtration chromatography in the presence of 2% octylglucoside, is consistent with a dimeric structure (15).

NSAID INHIBITION OF RECOMBINANT hPGHS-1 AND hPGHS-2

Two groups have reported that PGHS-1 and PGHS-2 can be distinguished on a pharmacologial basis since differences in their inhibition by classical NSAIDs have been detected (8,10). We have examined the ability of seven standard NSAIDs to inhibit the recombinant hPGHS isoforms either as purified enzymes or as crude microsomal preparations (7,14,16). For seven NSAIDs tested, including diclofenac, meclofenamic acid, indomethacin, sulindac sulfide, zomepirac, ibuprofen, and piroxicam, the rank orders of potencies against hPGHS-1 and -2 were the same (7). Although the IC_{50} values ranged from 0.1 µM to 300 µM, the IC_{50} values for diclofenac, meclofenamic acid, indomethacin, sulindac sulfide, and zomepirac against hPGHS-1 and hPGHS-2 demonstrated only up to a 3-fold selectivity for either enzyme. Our results are in contrast to that reported for murine PGHS-1 and PGHS-2 which display up to a 30-fold differential sensitivity to NSAID-mediated inhibition (8). The differences in the NSAID-mediated inhibition between the human and murine PGHS isoforms may reflect species-specific differences or differences in the methods of assaying NSAID inhibition of PGHS.

The method for assaying NSAID inhibition is critical as these drugs can be classified into two groups depending on their kinetic mechanism by which they inhibit PGHS (16). One class operates in a time independent manner, whereas the other demonstrates time dependent behavior involving a two step mechanism of inhibition (16). Time dependent NSAIDs are generally more potent since the second, time dependent step involves the formation of an essentially irreversibly inhibited enzyme-inhibitor complex.

We have investigated the kinetic mechanism of inhibition of purified hPGHS-1 and hPGHS-2 by four classical NSAIDs and a recently described NSAID, NS-398, which exhibits antiinflammatory activity in a number of animal models but is much less potent as an inducer of gastric lesions (9,12,13). In our studies the *in vitro* profile of NS-398 indicates that it is 260-fold selective for inhibition of hPGHS-2 over hPGHS-1 in terms of IC_{50} values (16). NS-398 and other classical NSAIDs, including flurbiprofen, meclofenamic acid, and indomethacin act as time dependent inhibitors of hPGHS-2 (16). In contrast, NS-398 is a time independent inhibitor of hPGHS-1, consistent with the formation of a reversible enzyme-inhibitor complex (16). The high degree of selectivity of NS-398 for hPGHS-2 over hPGHS-1 results from the difference in the nature of the time dependency of inhibition of the two isoforms.

15-HETE SYNTHESIS BY hPGHS-1 AND hPGHS-2

The formation of 11-HETE and 15-HETE by a lipoxygenase activity of PGHS-1 and the aspirin-stimulated formation of 15-HETE by recombinant murine PGHS-2 has been reported previously (6,22). We have detected the formation of 15-HETE from both recombinant hPGHS-1 and hPGHS-2 (7). Acetylation of both hPGHS-1 and -2 by aspirin results in an inhibition of PG synthesis by both enzymes, but has a differential effect on their capacity to form 15-HETE (7). Preincubation of hPGHS-2 with aspirin results in a 5-fold increase in the synthesis of 15-HETE; aspirin pretreatment of hPGHS-1 has no effect on its 15-HETE synthetic capacity. We have examined the enantiomeric composition of the 15-HETE produced by aspirin-treated hPGHS-2 using chiral phase HPLC and shown it to be entirely of the 15-*R*-HETE isomer as opposed to the 15-*S*-HETE isomer formed by lipoxygenases (7).

Both the cyclooxygenase activities of PGHS-1 and PGHS-2 are inhibited irreversibly by aspirin through acetylation of a specific serine residue (6,23). Site-specific mutagenesis studies indicate that this conserved serine residue (ser529 in hPGHS-1 and ser516 in hPGHS-2) is not involved in catalysis but that acetylation of this residue creates bulk close to the active site, altering access of substrate to the activated tyrosyl radical that initiates the initial hydrogen abstraction (6,7,17,23).

In an attempt to mimic the effects of aspirin acetylation on hPGHS-2 activity we mutated the putative aspirin acetylation site of the enzyme (ser516) by site-directed mutagenesis to a methionine residue and assessed the effect of this substitution on enzyme activity and the sensitivity of this activity to aspirin (17). The mutant hPGHS-2(ser516met) synthesized approximately 90-fold lower concentrations of PGE_2 than native protein (17). In contrast, the mutant hPGHS-2(ser516met) protein synthesized approximately 20-fold higher concentrations of 15-HETE than the wild-type protein and 1.7-fold higher concentrations of 15-HETE than aspirin acetylated hPGHS-2. Furthermore, 15-HETE and PG synthesis by hPGHS-2(ser516met) was unaltered by preincubation with aspirin, but was completely inhibited by preincubation with indomethacin (17). These results suggest that the side chain of methionine may sterically mimic an acetylated serine at the active site of hPGHS-2, and provide a model to study the effect of aspirin acetylation on the structure and activity of hPGHS-2. The high level synthesis of 15-R-HETE by aspirin-acetylated hPGHS-2, which is mimicked by hPGHS-2(ser516met), is a feature which clearly distinguishes hPGHS-2 from hPGHS-1, and suggests that the active sites of these two isoforms of PGHS have distinct structural features.

The demonstration that indomethacin can inhibit 15-HETE synthesis by hPGHS-2(ser516met) and aspirin-acetylated hPGHS-2 indicates that NSAIDs can still be accomodated at the active site of hPGHS-2 following modification of ser516 (17). We have further investigated the inhibition of 15-HETE synthesis by hPGHS-2 following aspirin acetylation by a range of structurally diverse NSAIDs in order to provide insight into the binding interactions of various NSAIDs to hPGHS-2 (18). All of the NSAIDs tested, including sulindac sulfide, ibuprofen, flurbiprofen, carprofen, naproxen, indomethacin, meclofenamate, NS-398, and diclofenac, inhibited aspirin-stimulated 15-HETE synthesis by hPGHS-2 (18). However, the rank order of potencies of compounds for the inhibition of aspirin-stimulated 15-HETE synthesis was different from that determined for their inhibition of PGE_2 by native, nonacetylated hPGHS-2 (18). The reasons for these differences are unclear. X-ray crystallographic studies of sheep PGHS-1 predicts a channel through which arachidonic acid penetrates in order to bind at the active site (24). The aspirin acetylation site of sheep PGHS-1 is located in this channel (24). As NSAIDs are structurally diverse, they may bind to the enzyme at different sites within a corresponding channel in hPGHS-2. Thus, the addition of an acetyl group at ser516 of hPGHS-2 may have differential effects on inhibitor binding depending on the location of the inhibitor binding site relative to ser516 (18).

SUMMARY

The recombinant high level expression systems described here have provided sufficient amounts of hPGHS-1 and hPGHS-2 for their complete biochemical and pharmacological characterization. The products of the reactions catalyzed by hPGHS-1 and -2 have been determined, and in addition data on the interaction of classical NSAIDs with the hPGHS isoforms indicates that conventional NSAIDs do not show a high degree of selectivity for either enzyme. Recently developed NSAIDs, such as NS-398, do show selectivity for hPGHS-2 via a difference in the kinetic mechanism of inhibition for each PGHS isoform. The differential effect of aspirin on the 15-HETE synthetic capacity of PGHS-2 versus PGHS-1, which can be mimicked by a single amino acid substitution in hPGHS-2, suggests that the active sites of the two enzymes are structurally dissimilar, and provides an additional insight into the interaction of these enzymes with NSAIDs and their substrates. It will be of interest to determine whether hPGHS-2 selective inhibitors, such as NS-398, will show therapeutic advantages over conventional NSAIDs, which show little selectivity for either enzyme and are associated with gastric and renal damage.

REFERENCES

1. Kujubu DA, Fletcher BS, Varnum BC, Lim RW, Herschman HR. TIS10, a phorbol ester tumor promoter-inducible mRNA from Swiss 3T3 cells, encodes a novel prostaglandin synthase/cyclooxygenase homologue. J Biol Chem. 1991; 266:12866-12872.

2. O'Banion MK, Winn VD, Young DA. cDNA cloning and functional activity of a glucocorticoid-regulated inflammatory cyclooxygenase. Proc Natl Acad Sci USA. 1992; 89:4888-4892.

3. Sirois J, Simmons DL, Richards JS. Hormonal regulation of messenger ribonucleic acid encoding a novel isoform or prostaglandin endoperoxide H synthase in rat preovulatory follicles. Induction *in vivo* and *in vitro*. J Biol Chem. 1992; 267:11586-11592.

4. Funk CD, Funk LB, Kennedy ME, Pong AS, Fitzgerald GA. Human platelet/crythroleukemia cell prostaglandin G/H synthase: cDNA cloning, expression, and gene chromosomal assignment. FASEB J. 1991; 5:2304-2312.

5. Hla T, Neilson K. Human cyclooxygenase-2 cDNA. Proc Natl Acad Sci USA. 1992; 89:7384-7388.

6. Lecomte M, Laneuville O, Ji C, DeWitt DL, Smith WL. Acetylation of human prostaglandin endoperoxide synthase-2 (cyclooxygenase-2) by aspirin. J Biol Chem. 1994; 289:13207-13215.

7. O'Neill GP, Mancini JA, Kargman S, Yergey J, Kwan MY, Falgueyret J-P, Abramovitz M, Kennedy BP, Ouellet M, Cromlish W, Culp S, Evans JF, Ford-Hutchinson AW, Vickers PJ. Overexpression of prostaglandin G/H synthases-1 and -2 by recombinant vaccinia virus:inhibition by nonsteroidal anti-inflammatory drugs and biosynthesis of 15-hydroxyeicosatetraenoic acid. Mol Pharmacol. 1994; 45:245-254.

8. Meade EA, Smith WL, DeWitt DL. Differential inhibition of prostaglandin endoperoxide synthase (cyclooxygenase) isozymes by aspirin and other non-steroidal anti-inflammatory drugs. J Biol Chem. 1993; 268:6610-6614.

9. Masferrer JL, Zweifel BS, Manning PT, Hauser SD, Leahy KM, Smith WG, Isakson PC, Seibert K. Selective inhibition of inducible cyclooxygenase 2 *in vivo* is antiinflammatory and nonulcerogenic. Proc Natl Acad Sci USA. 1994; 91:3228-3232.

10. Mitchell JA, Akarasereenont P, Thiemermann C, Flower RJ, Vane JR. Selectivity of nonsteroidal antiinflammatory drugs as inhibitors of constitutive and inducible cyclooxygenase. Proc Natl Acad Sci USA. 1994; 90:11693-11697.

11. Vane JR, Mitchell JA, Appleton I, Tomlinson A, Bishop-Bailey D, Croxtall J, Willoughby DA. Inducible isoforms of cyclooxygenase and nitric-oxide synthase in inflammation. Proc Natl Acad Sci USA. 1994; 91:2046-2050.

12. Futaki N, Takahashi S, Yokoyama M, Arai I, Higuchi S, Otomo S. NS-398, a new anti-inflammatory agent, selectively inhibits prostaglandin G/H synthase/cyclooxygenase (COX-2) activity in vitro. Prostaglandins. 1994; 47:55-59.

13. Futaki N, Yoshikawa K, Hamasaka Y, Arai I, Higuchi S, Iizuka H, Otomo S. NS-398, a novel non-steroidal anti-inflammatory drug with potent analgesic and antipyretic effects, which causes minimal stomach lesions. Gen Pharmac. 1993; 24:105-110.

14. Cromlish WA, Payette P, Culp SA, Ouellet M, Percival MD, Kennedy BP. High level expression of active human cyclooxygenase-2 in insect cells. Arch Biochem Biophys. 1994; (in press)

15. Percival MD, Ouellet M, Vincent C, Yergey J, Kennedy BP, O'Neill GP. Purification and characterization of recombinant human cyclooxygenase-2. (submitted)

16. Ouellet M, Percival MD. Effect of inhibitor time dependency on selectivity towards cyclooxygenase isoforms-1 and 2 by NS-398 and classic anti-inflammatory drugs. (submitted)

17. Mancini JA, O'Neill GP, Bayly C, Vickers P. Mutation of serine-516 in human prostaglandin G/H synthase-2 to methionine or aspirin acetylation of this residue stimulates 15-R-HETE synthesis. FEBS Lett. 1994; 342:33-37.

18. Vickers PJ, Boily C, Mancini JA, O'Neill GP. Non-steroidal anti-inflammatory drugs inhibit aspirin-stimulated 15-R-HETE synthesis by human prostaglandin G/H synthase-2. (submitted)

19. Shimokawa T, Smith WL. Expression of prostaglandin endoperoxide synthase-1 in a baculovirus system. Biochem Biophys Res Commun. 1992; 183:975-982.

20. Otto JC, DeWitt DL, Smith WL. N-glycosylation of prostaglandin endoperoxide synthases-1 and -2 and their orientations in the endoplasmic reticulum. J Biol Chem. 1993; 268:18234-18242.

21. Moss B. Vaccinia virus: a tool for research and vaccine development. Science. 1991; 252:1662-1667.

22. Hemler ME, Crawford CG, Lands WEM. Lipoxygenation activity of purified prostaglandin-forming cyclooxygenase. Biochemistry. 1978; 17:1772-1779.

23. DeWitt DL, El-Harith EA, Kraemer SA, Andrews MJ, Yao EF, Armstrong RL, Smith WL. The aspirin and heme-binding sites of ovine and murine prostaglandin endoperoxide synthases. J Biol Chem. 1990; 265:5192-5198.

24. Picot D, Loll PJ, Garavito RM. The X-ray crystal structure of the membrane protein prostaglandin H$_2$ synthase-1. Nature. 1994; 367:243-249.

AAS 46
Novel Molecular Approaches
to Anti-Inflammatory Theory
© 1995 Birkhäuser Verlag Basel

NEW FRONTIERS IN THE PHARMACOTHERAPY OF INFLAMMATION

Roderick John Flower

Department of Biochemical Pharmacology, The William Harvey Research Institute, The Medical College of St. Bartholomew's Hospital, Charterhouse Square, London EC1M 6BQ, UK.

Asking a pharmacologist to define new frontiers in the pharmacotherapy of inflammation is analogous to asking an astronomer which new star systems he will discover when the next space telescope is launched! It is an inherent property of research work that one can not foresee the discoveries of tomorrow although of course one can speculate as to the inevitable outcome of the trends which are visible today.

Even in science it is impossible to escape the conditioning effect of the past and it is sometimes a useful, as well as a sobering exercise to consider what our predecessors both in the recent and distant past have to say about inflammation. Such reflections may often cause us mirth but also, on occasions, provoke a new lines of thought or at least force us to a reassessment of our current belief systems. In science, the frontiers are never stationary but are constantly changing and moving. Let us look at where these boundaries used to be in the past before speculating about where they may lie in the future.

Some ancient ideas about inflammation

Our forebears were well conversant with the notion of inflammation and inflammatory disease. An Egyptian hieroglyph allegedly depicting the latter is thought to be derived from the hieroglyph for "fire" and this similarity, between heat and inflammation, seems to have been one which impressed itself strongly upon the minds of ancient scientists and physicians (reviewed in ref 1) . For example, in the Greek Hippocratic writings the term *phlegmone* - fiery heat - was used when describing disorders thought to be due to an influx of blood into a normally "bloodless" area. Indeed, the very name "inflammation" is derived from the Latin word *flamma* for fire.

The notion that inflammation is bought about by blood entering an area where it was not supposed to be is also one which appears in the writings of many ancient physicians such as Erasistratus (250 BC) and later, Galen (129-200AD), Pardoux (1641) and Boerhaave (1663-1738). In some cases it was believed that, in filling the vascular tissue, the new blood forced out other substances such as *pneuma* (Erasistratus) or that the blood itself mixed with substances such as *phlegm* giving rise to oedema or with *black bile* thus giving rise to cancers (Galen). Others believed that the blood in some way caused an obstruction or *acrimony* whereby damage was sustained to the blood vessels (Boerhaave).

Such ideas seem bizarre to us but it must not be thought that these ancient physicians had nothing of interest to say, for whilst their explanation of disease phenomena seem far fetched their clinical descriptions of inflammatory disorders were often highly accurate. It was in fact a Roman medical encyclopaedist, Celsus, living in 1 A.D. who first coined the phrase "heat, swelling, redness and pain" to describe the cardinal signs of inflammation - a phrase which will be found, in many cases unmodified, in a great many textbooks of pathology.

It was probably the English surgeon and naturalist John Hunter who provided us with the first really useful insight into the inflammatory process itself (2). His view was that inflammation was the response of the body to a disease or injury and not a disease in itself. He believed that inflammation was inseparable in many ways from the healing process. Like many of his predecessors he saw extravasation of blood from the vascular tree as the initiating stimulus and accorded the blood vessels a central role in the pathogenesis of the inflammation. The Hunterian notion of the essential similarity of inflammation and healing is one to which we will return later for I think that it is a point of view which has a lot to commend it.

Although his contemporaries and immediate successors were not always in agreement with his views, Hunter's theory about inflammation seems to mark a sort of watershed in the development of thinking about the inflammatory response. Another major step forward was made by people such as Addison (1802-1881) and Waller (1816-1870) when they identified the importance of white corpuscles in inflammation - work which was eventually developed so successfully by Virchow (1871) and later by Metchnikoff (1892) in their work on phagocytosis (1892).

Modern Developments in Inflammation

Perhaps no one idea has driven inflammation research more unrelentingly than the "mediator theory". We owe the concept, that the signs and symptoms of inflammation are caused by the release of chemicals, to the developments in inflammation research which took place in the beginning of the 20th century. This idea probably had its origins in the discovery, by immunologists, of soluble antibodies as mediators of immunological phenomena. The concept found a champion in Sir Henry Dale who was responsible for revolutionising our thinking about many physiological processes. In defining the term "auto pharmacology" that he coined (cf 3), Dale referred to an idea which grew out his work on synaptic transmission, namely that many effects, both physiological and pathophysiological, are bought about by the release of chemical mediators of which acetylcholine was, at the time, the most notable. In seeking to explain anaphylaxis and other inflammation - related phenomena, Dale invoked the release of what was then a new substance, histamine, which may therefore rightly claim to be the first inflammatory mediator.

The history of histamine research is an interesting one which cannot be covered in detail here. Suffice to say, that Dale's original ideas about the involvement of histamine were in part speculation and were based upon his observation that the substance could reproduce many of the effects of anaphylaxis *in vivo* (4). At that time, histamine had not been identified as a normal constituent of tissues, in fact it had first appeared as a chemical synthesised for other reasons by Windaus and Vogt in 1907 (5). Some years later the substance was extracted from putrefying protein mixtures by two groups and subsequently found as a contaminant in preparation of ergot by Dale and his colleague Barger, hence his interest in the substance. It was not until 1927 that the presence of histamine in mammalian tissues was definitively demonstrated by his group (6).

The evidence that histamine was *the* mediator of anaphylaxis and of the inflammatory response seemed very compelling and in reading contemporary accounts of this era we can catch echoes of what must have been the great disappointment felt by many workers when the first selective anti-histamines (such as mepyramine), whilst active in many models, were found not to be effective in diseases such as asthma. Despite this set back, the notion that one mediator might be responsible for the vast majority of inflammatory signs and symptoms has continued to haunt researchers. I call this the "Holy Grail hypothesis" for the discovery of such a mediator is indeed a noble aim and yet, like the Grail itself, has proved to be an elusive quarry.

Since histamine we have witnessed the discovery and testing of a great many other mediators. Serotonin followed close upon the heels of histamine and following this, an investigation of the pharmacological effects of trypsin and snake venoms lead to the discovery of the first peptide mediator, bradykinin (reviewed in ref 7). Many other "mediators" of various sorts were described in the late 30's and early 40's including extracts from inflammatory exudates, but often these substances were crude mixtures and no single active principle was convincingly isolated.

Few investigators working at the time of Sir Henry Dale would have anticipated the emergence of an entirely new group of mediators derived from lipids. Indeed, it is almost true to say that little significance was attached to lipids by the majority of biologists before 1930. They were regarded as a heterologous collection of "greasy" substances which could be extracted from animal and plant tissues by treatment with organic solvents such as ether, benzene or chloroform. Lipids were, it was recognised, structural components of our tissues: in adipose tissue the neutral lipids represented a useful metabolic store: in the diet, fat, was a convenient though not essential source of nutritional energy. That we now know this idea to be untrue is entirely due to a sequence of seminal though apparently unrelated discoveries made in the 1930's which began with the work of George Burr through which the concept of "essential fatty acids" was developed through to the observations of Von Euler (who incidentally worked with Dale) and which culminated in the first description of the prostaglandins (reviewed in 8).

Unlike the situation with histamine, where the scientific community had to await the synthesis of antagonists with which to validate and test their ideas, an astonishing finding made by Vane and his colleagues in the early 1970's (9,10, 11,12) pointed to the fact that not only did we already possess inhibitors of prostaglandin formation in our pharmacological armarmentarium but furthermore, a large number of them were in constant (empirical) clinical use already for the treatment of inflammatory conditions and the relief of pain and fever. This realisation immediately promoted prostaglandins to the top of the league table of likely candidates for The Grail. After all, what could be more compelling evidence for their primacy than the knowledge that they supplied the key to the mechanism of action of a group of drugs which had been around in one form or other since the beginning of the century?

But other surprises and revelations were to follow: the discovery of the structure of SRS-A and its identification as a member (or mixture of members) of the leukotriene family also derived from arachidonic acid was another major breakthrough (13) and, together with the elucidation of the lipoxygenase pathway, provided further evidence that the lipids were of cardinal importance in the inflammatory response. Confirming the ascendancy of lipids at that time was the discovery of platelet activating factor (14) another derivative of phospholipid metabolism with profound effects on certain aspects of the inflammatory response, especially in airways.

To a large extent progress in biology is dependent upon, and is underpinned by, advances in analytical methodology and techniques. It is to bioassay that we owe the discovery of substances such as histamine, bradykinin, the prostaglandins, SRS-A etc but it is to the chemist we turn to for structural information. When would we be if (to take one example) the mass spectrometer had never been invented and had not been so brilliantly exploited by the Karolinska group for the elucidation of the structures of the prostaglandins and leukotrienes? Today's technology is of course based firmly upon molecular biology and it is with the aid of these tools that we have been able to search out and analyse the latest wave of mediators, the cytokines. To the inflammation watcher, the discovery of cytokines and cytokine "networks" is analogous to the discovery of a huge underground edifice which, unbeknown to us, had underpinned all our theoretical structures in the inflammation field. Through their action on the genome, pluripotent agents such as TNF and IL-1 are able to organise, recruit and modulate so many features of the inflammatory response (reviewed in ref 15) that they must surely come close to. The Grail of inflammation research. This area is still under intensive research and it will be a while before its true promise is fully realised. In the meantime let us glance at another issue.

Is inflammation inevitable?

There is a tendency in our community to view inflammation as something which inevitably follows an injury or an infection: the idea seems to be that inflammation is something which "happens to you" and that once it has been initiated it can only be stopped by drug treatment. This is reinforced by the vast majority of the work which has focused upon mechanisms *leading to* mediator generation. However we must never forget that many if not most inflammatory episodes heal or indeed never progress beyond a very early phase. Several recent studies highlight this phenomenon in a dramatic way. For example in the experiments of Sternberg *et al* (16) it was observed that two histocompatible strains of rat behaved differently following injection of streptococcal cell wall extract: in one group of rats, of the Lewis strain, the injection of the antigen lead to profound polyarthritis but in the histocompatible Fischer strain only a transient inflammation was seen. This simple but compelling experiment demonstrates that it is not sufficient to have the injurious stimulus present and that there must be powerful counter-inflammatory processes latent within the body. So what was the difference between the two strains? Sternberg and her associates pinpointed the HPA axis and demonstrated that the resistant strain of rat responded to the injurious stimulus with a brisk release of anti-inflammatory corticosterone and ACTH whereas the Lewis strain did not.

This type of finding has been confirmed in other animal models and evidence has been provided by Chikanza *et al* (17) that a similar phenomenon may also obtain in man . In a cleverly designed experiment they compared the ability of patients with rheumatoid arthritis or osteoarthritis to respond to the stress of surgery, which was selected as a stimulus as both groups of patients required joint replacement. Once again it was found that those patients suffering from rheumatoid arthritis appeared to develop a less profound HPA axis response to the surgical trauma than those with osteoarthritis.

These experiments demonstrate to us that, under normal circumstances, the body is perfectly capable of moderating and dampening down the inflammatory process such that it may not even manifest at all, but that under some conditions, perhaps genetically determined, it is unable to bring about a resolution. A key element in this ability to counter inflammatory responses seems to lie within the HPA axis system which in itself is interesting as this axis is one of the main links between our perception of the world and our internal hormonal milieu.

I believe strongly that by studying the endogenous mechanisms which act to depress the inflammatory response, such as lipocortin 1 (18), we will discover new ways of supplementing these processes when they do not function properly in the chronically inflamed patient.

The development of anti-inflammatory drugs

The two major groups of drugs used in the treatment of inflammatory disease, the aspirin like drugs and the glucocorticoids have both come to us through the clever exploitation of an unusual observation or by serendipity followed up by systematic development. We live now in an era where it is thought proper to use "rational" processes to design drugs and the very choice of the word consigns all other approaches, by implication, to the realms of "anti-science". But this is not the case. All procedures ranging from empirical screening right the way through to molecular modelling have a rational basis although it is different in each case and whilst it is perfectly true that the design of an inhibitor using (say) a molecular modelling approach is much more intellectually satisfying than finding a "screening hit" by empirical means, there is surely room for both approaches in today's pharmaceutical industry. After all, it would take even the most powerful computer a long time to generate the structure of aspirin given that arachidonic acid was the substrate for the cyclooxygenase!

Somebody once remarked (and many a true word is spoken in jest) that the worst thing that ever happened in the field of inflammation research was the discovery of aspirin. The point about this comment is really that industry can become very market-orientated and this tends to counter innovation. A marketing department taking stock of the current prescriptions issued for the treatment of rheumatoid arthritis, would find it difficult to avoid the conclusion that in order to obtain a share of the market they would need a new non-steroidal, thus closing down any other options for research! Scientists can also be led into a different sort of trap. Once the anti-inflammatory activity of aspirin was observed in humans it was tested in animals (the reverse of the usual situation!). Convenient screens were devised which picked up the activity of aspirin (eg. the carrageenin paw oedema) which were then used to search for other drugs . This exercise, not surprisingly produced a crop of agents many of which acted through the same mechanism as aspirin. I am glad to say that the papers presented at this conference give us all every confidence that the basic research into the inflammatory process today is indeed truly innovative and that we are not being influenced too strongly by a market-lead approach.

Where will the next generation of anti-inflammatory drugs come from?

It is difficult for me to add anything to what has been said over the last two days at this meeting as it represents the cutting edge of research in our field. However, a few general comments might be in order.

It seems quite likely that the aspirin-prostaglandin story will reach its apotheosis with the discovery of selective Cox 2 inhibitors (see Seibert and O'Neill, this meeting). Although very little data is yet in the public domain it does seem as though such agents could offer all the advantages of a current aspirin like drugs coupled with none of the disadvantages associated with GI toxicity. If this were really to be achieved there is no doubt that the market for non-steroidal anti-inflammatory drugs would change dramatically with most of the old compounds being replaced by the new. It would lead to a rush to produce drugs which, even if they captured only a small portion of the huge market, could bring in substantial revenue. In taking lessons from the history of drug development in other areas, it does not necessarily follow that the first drug on the market will be the one which necessarily produces the most income in the long term.

In embarking upon a programme to discover selective Cox 2 inhibitors a pharmaceutical company begins with several advantages: first of all, the pharmacology of the aspirin like drugs is already very well known, secondly, both the Cox 1 and the Cox 2 enzymes have been sequenced and cloned and are available commercially and thirdly, the crystal structure of at least one of these enzymes is known thus encouraging the full use of powerful tools such as molecular modelling. It would be surprising, and profoundly disappointing, if selective Cox 2 inhibitors do not produce substantial changes in the treatment of patients with ongoing inflammatory disorders.

Like it or not, we are in an era where biologicals are important elements in our overall strategy for controlling severe forms of inflammatory disease. The identification of cytokine and other targets, and the advent of technology which allows us to "humanise" mouse monoclonal antibodies, has already lead to some very exciting preliminary data (19) and a qualitatively different type of anti-inflammatory profile to that seen with conventional agents. There remains, of course, the spectre of long term immunological problems whenever a foreign protein is injected, but the best guess at the moment is that at the very least these agents will be safe enough for the treatment of severe forms of inflammatory disorder.

Gene therapy for inflammation is still a way off and unfortunately our notions of what can be done in this area have outstripped the technology available to deliver target genes to their site of action, except in very few cases. One can see a future for this type of approach however: for example it is already possible to transfect the gene coding for the transmembrane regulatory element absent in cystic fibrosis in the nasal passages and the lungs of mice (20) and one might see the same type of technology being applied for the delivery of a gene which, for example, suppresses asthma in some way or other: perhaps an anti-sense sequence designed to correct the over-activity of an enzyme such as the 5-lipoxygenase. These are dreams at the moment but at the rate at which the field is progressing it would not surprise me to learn that by time this book is published some study or other is under way in this area!

Of course, not all inflammatory disorders are genetically linked. In fact the genetic basis, if any, of rheumatoid arthritis is far from clear but nevertheless I believe that the application of molecular biological techniques will make a big impact in *diagnostics* in the inflammation field in the forthcoming years. In the hypertension area for example a recent paper (21) described a strong association between the angiotensigen gene and hypertension in a closed community. If the genes which correlate strongly with lupus or rheumatoid disease could be identified it should be possible to predict very early on in the life time of a child whether or not he or she is likely to suffer from the disease and, even which drugs might be the most appropriate to use if should the disease manifest. In this way the onset of the disease can be checked much earlier and thereby improving the quality of life for sufferers.

It is from basic science and an investigation of basic mechanisms in inflammation, that virtually all future important therapeutic advances will be made and there is no doubt that this activity should continue to occupy us as one of our main endeavours. Perhaps, in line with the Hunterian idea of inflammation and healing as opposite sides of the same coin, we should turn our attention from those processes which *initiate* inflammation to an examination of those which naturally *suppress* it such as anti-inflammatory cytokines (eg. IL-10). The endocrinology of inflammation and in particular the activity of the HPA axis in suppressing facets of the inflammatory response also deserves further attention. There is a great deal more work to be done in the field of glucocorticoid regulation of inflammation: this is a Pandora's box and so far we have only lifted the lid and squinted in. I am sure that many more discoveries of importance await us here in due course.

REFERENCES

1. Smith SE. Historical survey of definitions and concepts of inflammation. In: Handbook of Experimental Pharmacology, 50/1. eds. Vane JR, Ferreira SH. Springer-Verlag, Berlin, Heidelberg, New York, 1978: 1-4.

2. Jarcho S. John Hunter on inflammation. Amer J Cardiol 1970: 26: 615-618.

3. Dale HH. Some chemical factors in the control of the circulation. Lancet 1929: 1: 1233-1237 & 1285-1290.

4. Dale HH, Laidlaw PP. The physiological action of ß iminazolylethylamine. J Physiol (Lond) 1910: 41: 318-344.

5. Windaus A, Vogt W. Synthese des imidazolylethylamines. Ber dtsch chem Ges 1907: 40: 3691-3697.

6. Best CH, Dale HH, Duddley HW, Thorpe WV. The nature of the vasodilator consituents of certain tissue extracts. J Physiol (Lond) 1927: 62: 397-417.

7. Rocha E Silva M. A brief history of inflammation. In: Handbook of Experimental Pharmacology, 50/1, eds. Vane JR, Ferreira SH. Springer-Verlag, Berlin, Heidelberg, New York, 1978: 6-25.

8. Willis AL. The eicosanoids: an introduction and an overview. CRC Handbook of eicosanoids: prostaglandins and related lipids., Vol 1 part A. CRC Press Inc Boca Raton Florida, 1987: 3-46.

9. Vane JR. Inhibition of prostaglandin synthesis as a mechanism of action for aspirin like drugs. Nature (Lond). New Biol 1971: 231: 232-235.

10. Ferreira S.H, Moncada S, Vane JR. Indomethacine and aspirin abolish prostaglandin release from the spleen. Nature (Lond) New Biol 1971: 231: 237-239.

11. Smith JB, Willis AL. Aspirin selectively inhibits prostaglandin production in human platelets. Nature (Lond) New Biol. 1971: 231: 235-237.

12. Flower R, Gryglewski R, Herbacynska-Cedro K, Vane JR. Effects of anti-inflammatory drugs on prostaglandin biosynthesis. Nature (Lond) New Biol 1972: 238: 104-106.

13. Samuelson B. Leukotrienes: a novel group of compounds including SRS-A. Prog Lipid Res 1982: 20: 23-30.

14. Beneveniste J, Henson PM, Cochrane CG. Leukocyte dependent histamine release from rabbit platelets: the role of IgE, basaphils and platelet activating factor. J Exp Med 1972: 136: 1356-1377.

15. Henderson B, Blake S. Therapeutic potential of cytokine manipulation. TIPS 1992: 13: 145-152.

16. Sternberg EM, Hill JM, Chrousos GP, Kamilaris T, Listwak SJ, Gold PW, Wilder RL. Inflammatory mediator induced hypothalamic-pituitary-adrenal axis activation

is defective is streptococcal cell wall arthritis susceptible Lewis rats. Proc Natl Acad Sci USA 1989: 86: 2374-2378.

17.   Chikanza IC, Petrou P, Kingsley G, Chrousos G, Panayi S. Defective hypothalamic response to immune and inflammatory stimuli in patients with rheumatoid arthritis. Arth Rheum 1992: 35: 1281-1288.

18.   Flower RJ, Rothwell NJ. Lipocortin 1: cellular mechanisms and clinical relevance. TIPS 1994: 15: 71-76.

19.   Lockwood CM, Thiru S, Isaacs JD, Hale G, Waldmann H. Long term   remission of intractable systemic vasculitis with monoclonal antibody therapy. Lancet 1993: 1620-1622.

20.   Hyde SC, Gill DR, Higgins CF, Tresize AEO, MacVinish LJ, Cuthbert AW, Ratcliff R, Evans MJ, Colledge WH. Correction of the ion transport defect in cystic fibrosis transgenic mice by gene therapy. Nature 1993: 362: 250-255.

21.   Caulfield M, Lavender P, Farrall M, Munroe P, Lawson M, Turner P, AJL Clark. Linkage of the angiotensinogen gene to essential hypertension.   New Eng J Med 1994: 330: 1629-1633.

AGENTS AND ACTIONS SUPPLEMENTS

K. Brune, *University of Erlangen, Germany*
M.J. Parnham, *Bonn, Germany*

Agents and Actions Supplements (AAS) is a book series for rapid publication of the proceedings of symposia on topics of current interest in inflammation, allergy, related respiratory diseases, thrombosis and related fields. The series allows fast dissemination of surveys and specialized reports on, for example, research into the role of prostaglandins in inflammation and thrombosis, new trends in the treatment of rheumatoid arthritis, allergic reactions and asthma.

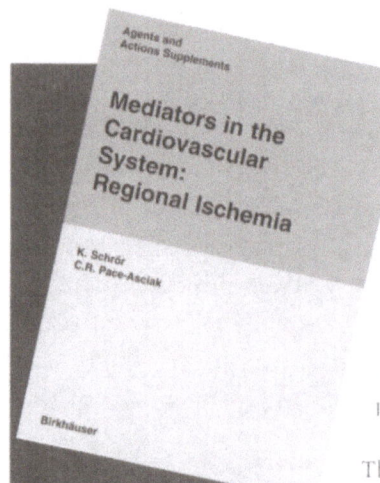

K. Schrör, *Heinrich-Heine University, Düsseldorf, Germany*
C.R. Pace-Asciak, *Hospital for Sick Children, Toronto, ONT, Canada (Eds)*

Mediators in the Cardiovascular System: Regional Ischemia

1995. 332 pages. Hardcover
ISBN 3-7643-5130-6 (AAS 45)

The study of chemical signalling between different cell types is one of the most exciting areas in current cardiovascular research. Significant progress has been made during the last few years and a number of intercellular mediators have been structurally identified and their regulation analysed. These developments have a major impact on cardiovascular pharmacology. This includes both the molecular design of new drugs, together with an improved understanding of the actions of established compounds.

Experts of international repute address in this volume relevant aspects of recent research and emerging themes in the field of myocardial ischemia. Particular emphasis is placed upon the regulation, function and pharmacological modification of eicosanoids, nitric oxide and endothelins, with an outlook on future drug developments.

Birkhäuser Verlag • Basel • Boston • Berlin

RESPIRATORY PHARMACOLOGY AND PHARMACOTHERAPY

D. Raeburn, *Rhône-Poulenc Rorer Ltd, Dagenham, UK*
M.A. Giembycz, *Royal Brompton National Heart and Lung Institute, London, UK (Eds)*

Airways Smooth Muscle: Biochemical Control of Contraction and Relaxation

1994. 352 pages. Hardcover
ISBN 3-7643-5043-1

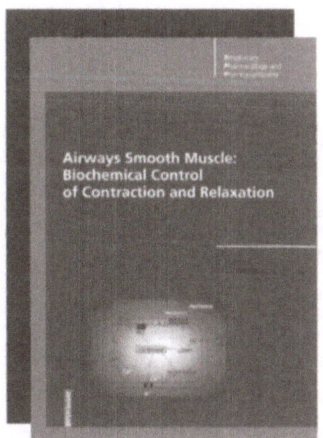

The study of airways smooth muscle has intensified over recent years against a background of a growing incidence of asthma and other respiratory disorders. Building on the previous volumes in the series, this research monograph focuses upon the biochemical regulation of contraction and relaxation of airways smooth muscle.

Written by international experts, this up-to-date reference work includes chapters on actin, myosin, diglyceride and protein kinase C, inositol polyphosphates, current theories regarding mechanisms of force generation and maintenance, G-proteins, cyclic nucleotides and properties of airways smooth muscle cells in culture.

All academic and clinical research workers in the field of airways smooth muscle physiology, biochemistry, pharmacology and cell and molecular biology will find this volume an indispensable source of information.

Birkhäuser Verlag • Basel • Boston • Berlin

RESPIRATORY PHARMACOLOGY AND PHARMACOTHERAPY

D. Raeburn, *Rhône-Poulenc Rorer Ltd, Dagenham, UK*
M.A. Giembycz, *Royal Brompton National Heart and Lung Institute, London, UK (Eds)*

Airways Smooth Muscle: Development, and Regulation of Contractility

1994. 420 pages. Hardcover
ISBN 3-7643-5011-3

The focus of this second volume in a new series of research mono-graphs is on the growth and development of airways smooth muscle, and its regulation. It also addresses the role of nerves and other physi-ological factors responsible for regulating contractility.

Internationally acclaimed experts review the latest research data and emerging themes in the field. Aspects discussed include trophic factors and the control of smooth muscle development, cell-to-cell coupling, electrophysiology, voltage-dependent calcium channels, and the effects of ageing on contractility.

This comprehensive and up-to-date work of reference is a valuable source of information which will benefit researchers in physiology, pharmacology, anatomy and developmental biology as well as clini-cians.

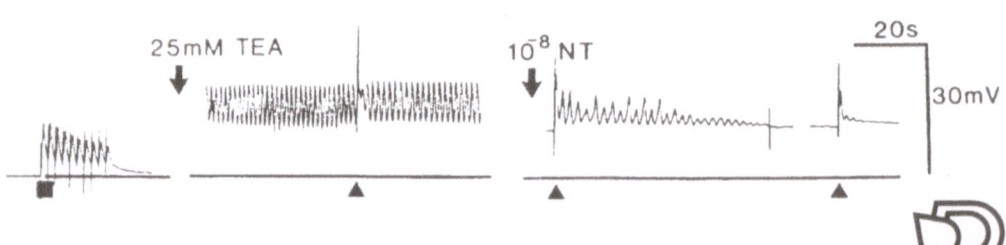

Birkhäuser Verlag • Basel • Boston • Berlin

D. Raeburn, *Rhône-Poulenc Rorer Ltd, Dagenham, UK*
M.A. Giembycz, *Royal Brompton National Heart and Lung Institute, London, UK (Eds)*

Airways Smooth Muscle: Structure, Innervation and Neurotransmission

1994. 328 pages. Hardcover
ISBN 3-7643-5010-5

In the context of a growing incidence of respiratory disorders worldwide, particularly asthma and allergy, the importance of research into the respiratory system is increasingly recognized. The emphasis of this first volume in a new series of research monographs is on the anatomical aspects of airways smooth muscle, including its innervation and neurotransmission.

Scientists of international repute were invited to contribute chapters on anatomy, gross morphology and ultrastructure, sympathetic, parasympathetic and NANC innervation, vagal reflexes, prejunctional regulation of neurotransmission, and neural elements in airways smooth muscle.

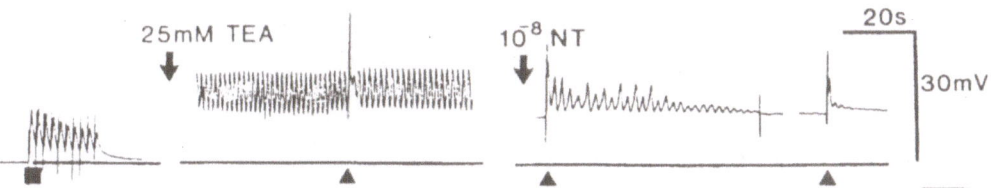

Birkhäuser Verlag • Basel • Boston • Berlin